Geological Structures and Maps

A PRACTICAL GUIDE

Related Pergamon titles

ANDERSON
Field Geology in the British Isles

ANDERSON
Structure of Western Europe

ANDERSON
Structure of the British Isles

CONDIE
Plate Tectonics and Crustal Evolution 3 ed

LISLE
Geological Strain Analysis

OWEN
The Geological Evolution of the British Isles

ROBERTS
Introduction to Geological Maps and Structures

Related Pergamon Journals

International Journal of Rock Mechanics and Mining Sciences & Geomechanics Abstracts

Journal of Structural Geology

Geological Structures and Maps

A PRACTICAL GUIDE

by

RICHARD J. LISLE

University College of Swansea, UK

PERGAMON PRESS

OXFORD · NEW YORK · BEIJING · FRANKFURT
SÃO PAULO · SYDNEY · TOKYO · TORONTO

U.K.	Pergamon Press plc, Headington Hill Hall, Oxford OX3 0BW, England
U.S.A.	Pergamon Press, Inc., Maxwell House, Fairview Park, Elmsford, New York 10523, U.S.A.
PEOPLE'S REPUBLIC OF CHINA	Pergamon Press, Room 4037, Qianmen Hotel, Beijing, People's Republic of China
FEDERAL REPUBLIC OF GERMANY	Pergamon Press GmbH, Hammerweg 6, D-6242 Kronberg, Federal Republic of Germany
BRAZIL	Pergamon Editora Ltda, Rua Eça de Queiros, 346, CEP 04011, Paraiso, São Paulo, Brazil
AUSTRALIA	Pergamon Press Australia Pty Ltd., P. O. Box 544, Potts Point, N.S.W. 2011, Australia
JAPAN	Pergamon Press, 5th Floor, Matsuoka Central Building, 1-7-1 Nishishinjuku, Shinjuku-ku, Tokyo 160, Japan
CANADA	Pergamon Press Canada Ltd., Suite No. 271, 253 College Street, Toronto, Ontario, Canada M5T 1R5

First edition 1988

Library of Congress Cataloging-in-Publication Data

Lisle, Richard J.
Geological structures and maps.
Bibliography: p.
Includes index.
1. Geology, Structural—Maps. I. Title.
QE601.2.L57 1988 912'.15518 88-5807

British Library Cataloguing in Publication Data

Lisle, Richard J.
Geological structures and maps.
1. Geological features. Analysis. Use of geological maps
I. Title
551.8

ISBN 0-08-034854-8 Hardcover
ISBN 0-08-034853-X Flexicover

Printed in Great Britain by A. Wheaton & Co. Ltd., Exeter

Preface

GEOLOGICAL maps represent the expression on the earth's surface of the underlying geological structure. For this reason the ability to correctly interpret the relationships displayed on a geological map relies heavily on a knowledge of the basic principles of structural geology.

This book discusses, from first principles up to and including first-year undergraduate level, the morphology of the most important types of geological structures, and relates them to their manifestation on geological maps.

Although the treatment of structures is at an elementary level, care has been taken to define terms rigorously and in a way that is in keeping with current professional usage. All too often concepts such as 'asymmetrical fold', 'fold axis' and 'cylindrical fold' explained in first textbooks have to be re-learned 'correctly' at university level.

Photographs of structures in the field are included to emphasize the similarities between structures at outcrop scale and on the scale of the map. Ideally, actual fieldwork experience should be gained in parallel with this course.

The book is designed, as far as possible, to be read without tutorial help. Worked examples are given to assist with the solution of the exercises. Emphasis is placed throughout on developing the skill of three-dimensional visualization so important to the geologist.

In the choice of the maps for the exercises, an attempt has been made to steer a middle course between the artificial-looking idealized type of 'problem map' and real survey maps. The latter can initially overwhelm the student with the sheer amount of data presented. Many of the exercises are based closely on selected 'extracts' from actual maps.

I am grateful to Professor T. R. Owen who realized the need for a book with this scope and encouraged me to write it. Peter Henn and Catherine Shephard of Pergamon Books are thanked for their help and patience. Thanks are also due to Vivienne Jenkins and Wendy Johnson for providing secretarial help, and to my wife Ann for her support.

Contents

Geological Map Symbols

inclined strata, dip in degrees

horizontal strata

vertical strata

axial surface trace of antiform

axial surface trace of synform

fold hinge line, fold axis or other linear structure, plunge in degrees

inclined cleavage, dip in degrees

horizontal cleavage

vertical cleavage

geological boundary

fault line, mark on downthrow side

younging direction of beds

metamorphic aureole

1

Geological Maps

1.1 What are geological maps?

A geological map shows the distribution of various types of bedrock in an area. It usually consists of a topographic map (a map giving information about the form of the earth's surface) which is shaded, or coloured to show where different rock units occur at or just below the ground surface. Figure 1.1 shows a geological map of an area in the Cotswolds. It tells us for instance that clays form the bedrock at Childswickham and Broadway but if we move eastwards up the Cotswold escarpment to Broadway Hill we can find oolitic limestones. Lines on the map are drawn to show the boundaries between each of the rock units.

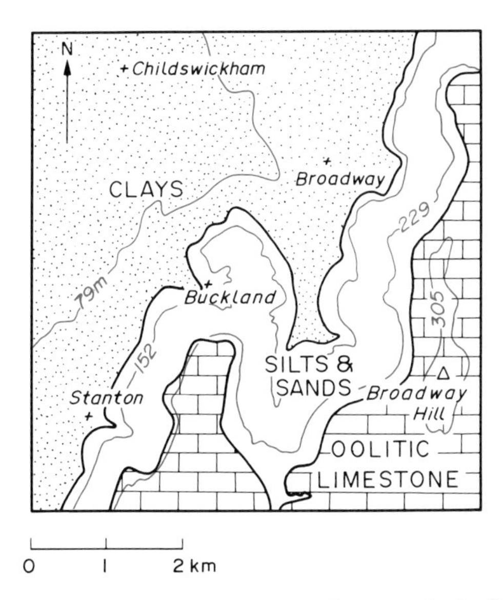

FIG. 1.1 A geological map of the Broadway area in the Cotswolds.

1.2 How is such a geological map made?

The geologist in the field firstly records the nature of rock where it is visible at the surface. Rock outcrops are examined and characteristics such as rock composition, internal structure and fossil content are recorded. By using these details, different units can be distinguished and shown separately on the base map. Of course, rocks are not always exposed at the surface. In fact, over much of the area in Fig. 1.1 rocks are covered by soil and by alluvial deposits laid down by recent rivers. Deducing the rock unit which underlies the areas of unexposed rock involves making use of additional data such as the type of soil, the land's surface forms (geomorphology) and information from boreholes. Geophysical methods allow certain physical properties of rocks (such as their magnetism and density) to be measured remotely, and are therefore useful for mapping rocks in poorly exposed regions. This additional information is taken into account when the geologist decides on the position of the boundaries of rock units to be drawn on the map. Nevertheless, there are always parts of the map where more uncertainty exists about the nature of the bedrock, and it is important for the reader of the map to realize that a good deal of interpretation is used in the map-making process.

1.3 What is a geological map used for?

The most obvious use of a geological map is to indicate the nature of the near-surface bedrock. This is clearly of great importance to civil engineers who, for example, have to advise on the excavation of road cuttings or on the siting of bridges; to geographers studying the use of land and to companies exploiting minerals. The experienced geologist can, however, extract more from the geological map. To the trained observer the features on a geological map reveal vital clues about the geological history of an area. Furthermore, the bands of colour on a geological map are the expression on the ground surface of layers or sheets of rock which extend and slant downwards into the crust of the earth. The often intricate pattern on a map, like the graininess of a polished wooden table top, provides tell-tale evidence of the structure of the layers beneath the surface. To make these deductions first requires knowledge of the characteristic form of common geological structures such as faults and folds.

This book provides a course in geological map reading. It familiarizes students with the important types of geological structures and enables them to recognize these as they would appear on a map or cross-section.

2

Uniformly Dipping Beds

2.1 Introduction

Those who have observed the scenery in Western movies filmed on the Colorado Plateau will have been impressed by the layered nature of the rock displayed in the mountainsides. The layered structure results from the deposition of sediments in sheets or beds which have large areal extent compared to their thickness. When more beds of sediment are laid down on top the structure comes to resemble a sandwich or a pile of pages in a book (Fig. 2.1). This stratified structure is known as bedding.

FIG. 2.1 Horizontal bedding. **A**: Lower Jurassic, near Cardiff.

FIG. 2.1 Horizontal bedding. **B**: Upper Carboniferous, Cornwall

In some areas the sediments exposed on the surface of the earth still show their unmodified sedimentary structure; that is, the bedding is still approximately horizontal. In other parts of the world, especially those in ancient mountain belts, the structure of the layering is dominated by the buckling of the strata into corrugations or folds so that the slope of the bedding varies from place to place. Folds, which are these crumples of the crust's layering, together with faults where the beds are broken and shifted, are examples of geological structures to be dealt with in later chapters. In this chapter we consider the structure consisting of planar beds with a uniform slope brought about by the tilting of originally horizontal sedimentary rocks.

2.2 Dip

Bedding and other geological layers and planes which are not horizontal are said to dip. Figure 2.2 shows field examples of dipping beds. The *dip* is the slope of a geological surface. There are two aspects to the dip of a plane:

FIG. 2.2 Dipping beds in Teruel Province, Spain. **A**: Cretaceous Limestones dipping at about 80°. **B:** Tertiary conglomerates and sandstones dipping at about 50°.

(a) *the direction of dip*, which is the compass direction towards which the plane slopes; and

(b) the *angle of dip*, which is the angle that the plane makes with a horizontal plane (Fig. 2.3).

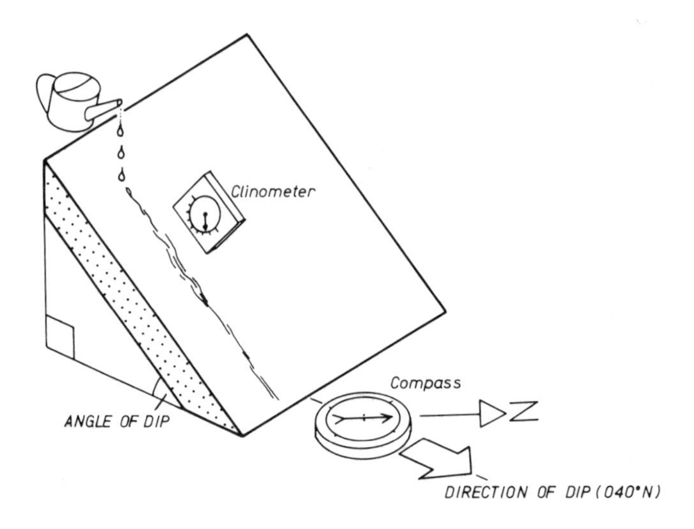

FIG. 2.3 The concepts of direction of dip and angle of dip.

The direction of dip can be visualized as the direction in which water would flow if poured onto the plane. The angle of dip is an angle between 0° (for horizontal planes) and 90° (for vertical planes). To record the dip of a plane all that is needed are two numbers; the direction of dip followed by the angle of dip, e.g. 138/74 is a plane which dips 74° in the direction 138°N (this is a direction which is SE, 138° clockwise from north). In the field the direction of dip is usually measured with a magnetic compass which incorporates a device called a clinometer, based on a plumbline or spirit level principle, for the measurement of dip angles.

2.3 Plunge of lines

With the help of Fig. 2.4 imagine a dipping plane with a number of straight lines drawn on it in different directions. All these lines are said to be contained within the plane and are parallel to the plane. With the exception of line 5 the lines are not horizontal; we say they are plunging lines. Line 5 is non-plunging. *Plunge* is used to describe the tilt of lines, the word dip being reserved for planes. The plunge fully expresses the tilt of a line and has two parts:

(a) the plunge direction, and
(b) the angle of plunge.

Consider the vertical plane which passes through the plunging line in Fig. 2.5. The *plunge direction* is the direction in which this vertical plane runs, and is the direction towards which the line is tilted. The *angle of plunge* is the

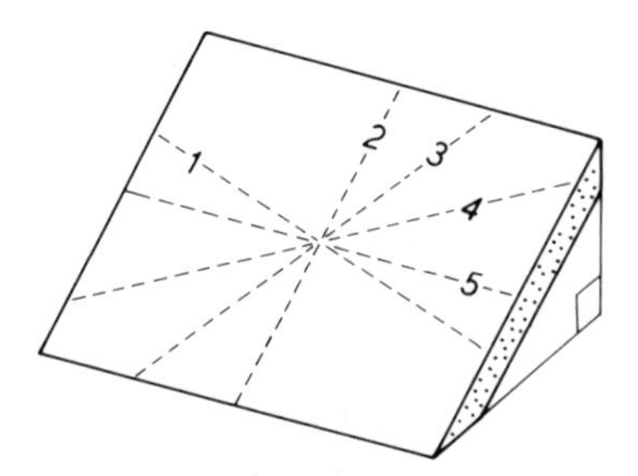

FIG. 2.4 Lines geometrically contained within a dipping plane.

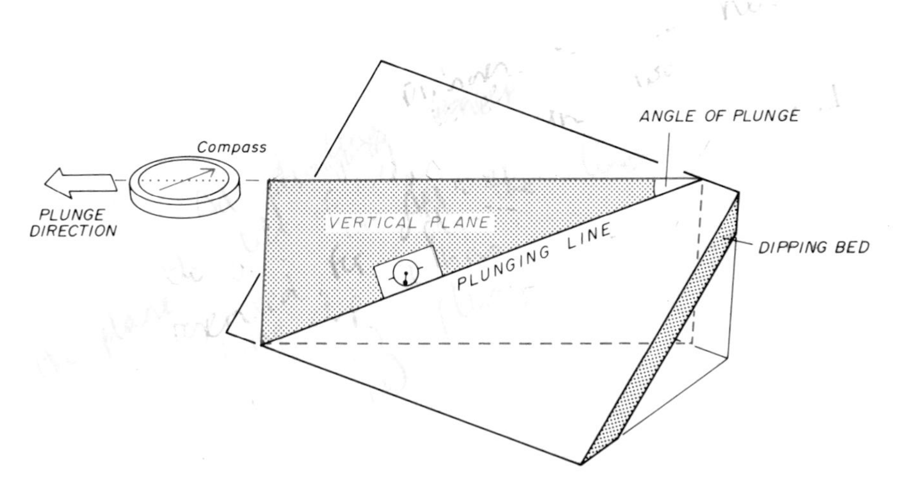

FIG. 2.5 The concepts of direction of plunge and angle of plunge.

amount of tilt; it is the angle, measured in the vertical plane, that the plunging line makes with the horizontal. The angle of plunge of a horizontal line is 0° and the angle of plunge of a vertical line is 90°. The plunge of a line can be written as a single expression, e.g. 325-62 describes a line which plunges 62° towards the direction 325°N. So far we have illustrated the concept of plunge using lines drawn on a dipping plane but, as we shall see later, there are a variety of linear structures in rocks to which the concept of plunge can be applied.

2.4 Strike lines

Any dipping plane can be thought of as containing a large number of lines of varying plunge (Fig. 2.4). The *strike line* is a non-plunging or horizontal line within a dipping plane. The line numbered 5 in Fig. 2.4 is an example of a strike line; it is not the only one but the other strike lines are all parallel to it. If we think of the sloping roof of a house as a dipping plane, the lines of the ridge and the eaves are equivalent to strike lines.

Within the dipping plane the line at right angles to the strike line is the line with the steepest plunge. Verify this for yourself by tilting a book on a flat tabletop as shown in Fig. 2.6. Place a pencil on the book in various orientations. The plunge of the pencil will be steepest when it is at right angles to the spine of the book (a strike line). The angle of plunge of the steepest plunging line in a plane is equal to the angle of dip of that plane.

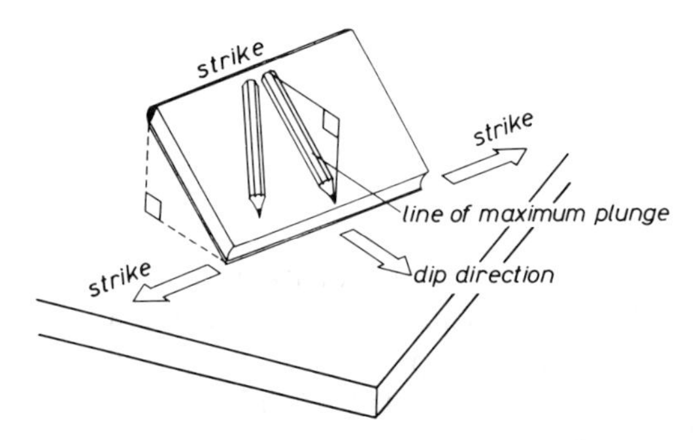

FIG. 2.6 A classroom demonstration of a dipping plane.

When specifying the direction of a strike line we can quote either of two directions which are 180° different (Fig. 2.6). For example, a strike direction of 060° is the same as a strike direction of 240°. The direction of dip is always at right angles to the strike and can therefore be obtained by either adding or subtracting 90° from the strike whichever gives the down-dip direction.

The map symbol used to represent the dip of bedding usually consists of a stripe in the direction of the strike with a short dash on the side towards the dip direction, (see list of symbols at the beginning of the book). Some older maps display dip with an arrow which points to the dip direction.

2.5 Apparent dip

At many outcrops where dipping beds are exposed the bedding planes themselves are not visible as surfaces. Cliffs, quarries and cuttings may provide more or less vertical outcrop surfaces which make an arbitrary angle with the strike of the beds (Fig. 2.7A). When such vertical sections are not perpendicular to the strike (Fig. 2.7B), the beds will appear to dip at a gentler angle than the true dip. This is an *apparent dip*.

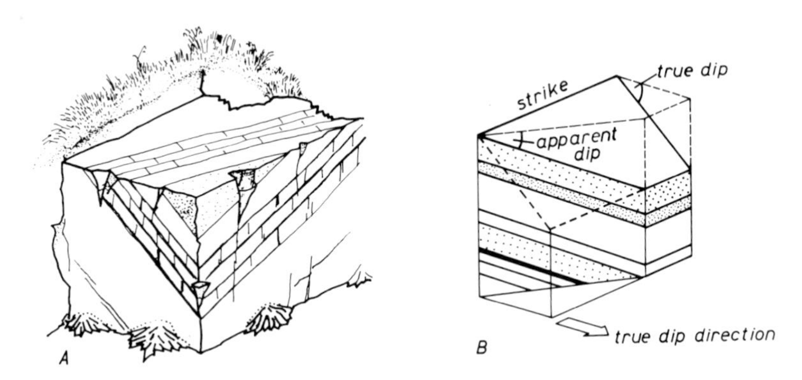

FIG. 2.7 Apparent dip.

It is a simple matter to derive an equation which expresses how the size of the angle of apparent dip depends on the true dip and the direction of the vertical plane on which the apparent dip is observed (the section plane). In Figure 2.8 the obliquity angle is the angle between the trend of the vertical section plane and the dip direction of the beds.

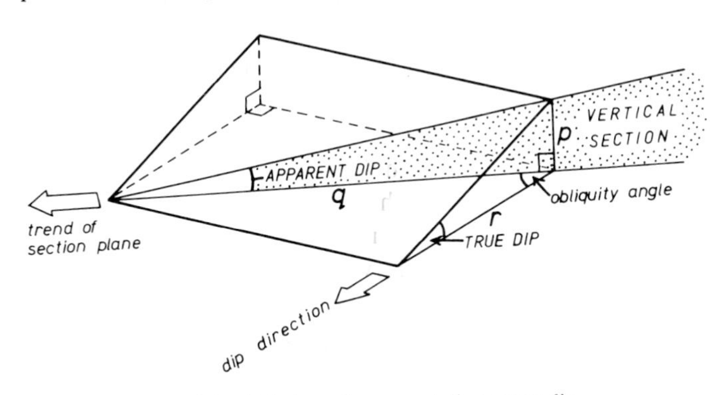

FIG. 2.8 Relation of apparent dip to true dip.

From Fig. 2.8:

Tangent (apparent dip) $= p/q$,

Tangent (true dip) $= p/r$ and

Cosine (obliquity angle) $= r/q$.

Therefore:

Tan (apparent dip) = *Tan* (true dip) × *Cos* (obliquity angle).

It is sometimes necessary to calculate the angle of apparent dip, for instance when we want to draw a cross-section through beds whose dip direction is not parallel to the section line.

2.6 Outcrop patterns of uniformly dipping beds

The geological map in Fig. 2.9A shows the areal distribution of two rock formations. The line on the map separating the formations has an irregular shape even though the contact between the formations is a planar surface (Fig. 2.9B).

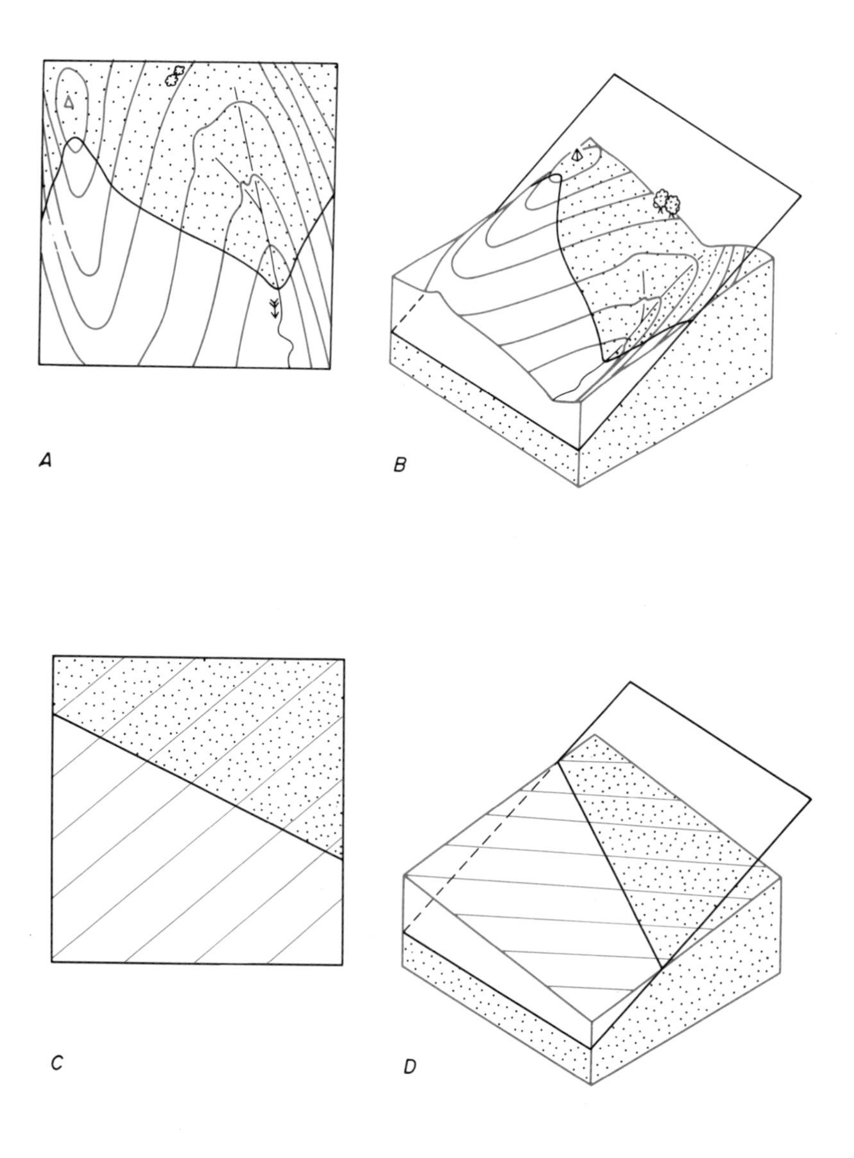

FIG. 2.9 The concept of outcrop of a geological contact.

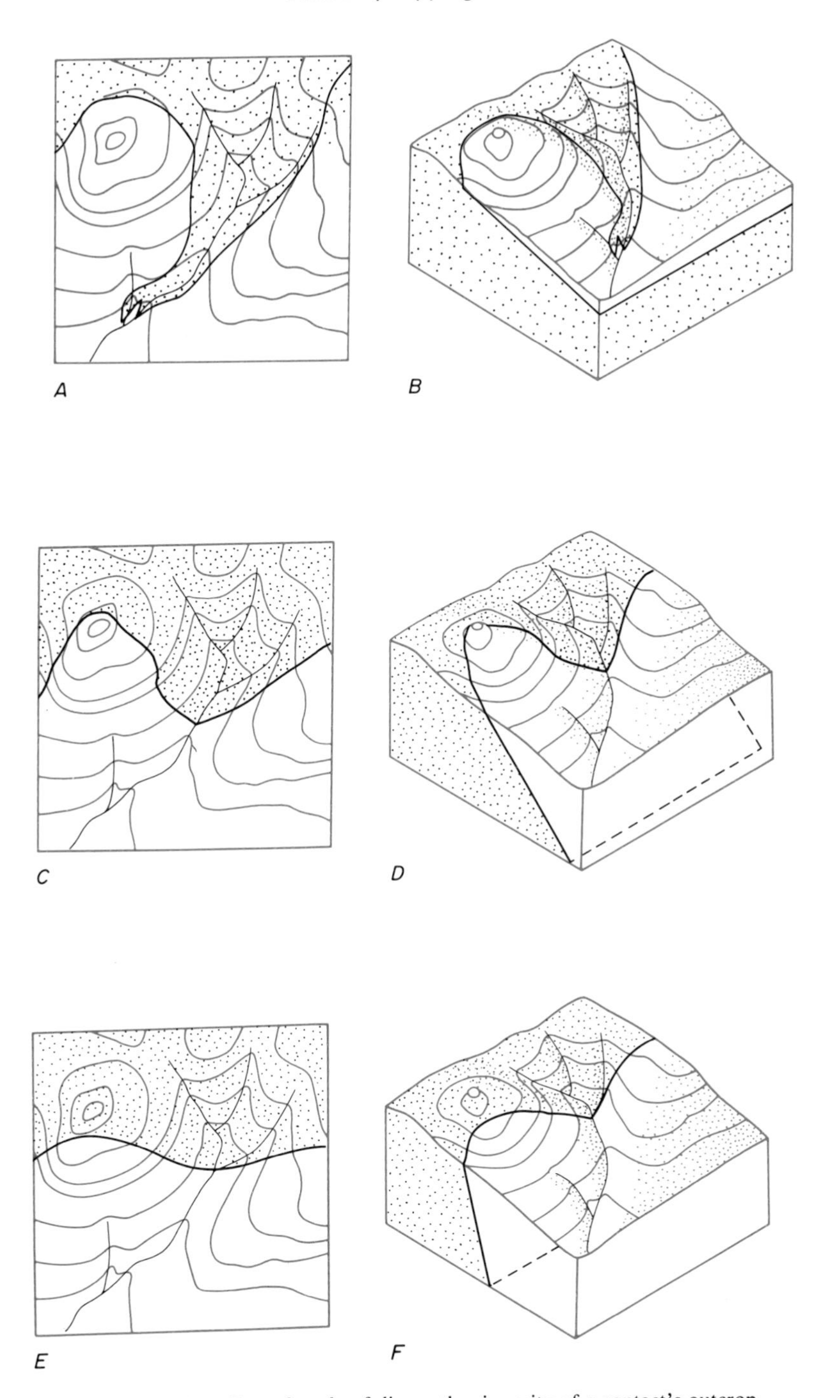

FIG. 2.10 The effect of angle of dip on the sinuosity of a contact's outcrop.

To understand the shapes described by the boundaries of formations on geological maps it is important to realize that they represent a line (horizontal, plunging or curved) produced by the intersection in three dimensions of two surfaces (Fig. 2.9B, D). One of these surfaces is the 'geological surface'—in this example the surface of contact between the two formations. The other is the 'topographic surface'—the surface of the ground. The topographic surface is not planar but has features such as hills, valleys and ridges. As the block diagram in Fig. 2.9B shows, it is these irregularities or topographic features which produce the sinuous trace of geological contacts we observe on maps. If, for example, the ground surface were planar (Fig. 2.9D), the contacts would run as straight lines on the map (Fig. 2.9C).

The extent to which topography influences the form of contacts depends also on the angle of dip of the beds. Where beds dip at a gentle angle, valleys and ridges produce pronounced 'meanders' (Fig. 2.10A, B). Where beds dip steeply the course of the contact is straighter on the map (Fig. 2.10C, D, E, F). When contacts are vertical their course on the map will be a straight line following the direction of the strike of the contact.

2.7 Representing surfaces on maps

In the previous section two types of surface were mentioned: the geological or structural surface and the topographic or ground surface. It is possible to describe the form of either type on a map. The surface shown in Fig. 2.11A can be represented on a map if the heights of all points on the surface are specified on the map. This is usually done by stating, with a number, the elevation of individual points such as that of point X (a spot height) and by means of lines drawn on the map which join all points which share the same height (Fig. 2.11B). The latter are *contour lines* and are drawn usually for a fixed interval of height. Topographic maps depict the shape of the ground usually by means of *topographic contours*. *Structure contours* record the height of geological surfaces.

2.8 Properties of contour maps

Topographic contour patterns and structure contour patterns are interpreted in similar ways and can be discussed together. Contour patterns are readily understood if we consider the changing position of the coastline, if sea level were to rise in, say, 10-metre stages. The contour lines are analogous to the shoreline which after the first stage of inundation would link all points on the ground which are 10 metres above present sea level and so on. For a geological surface the structure contours are lines which are everywhere parallel to the local strike of the dipping surface. The local direction of slope (dip) at any point is at right angles to the trend of the contours. Contour lines will be closer together when the slope (dip) is steep. A uniformly sloping

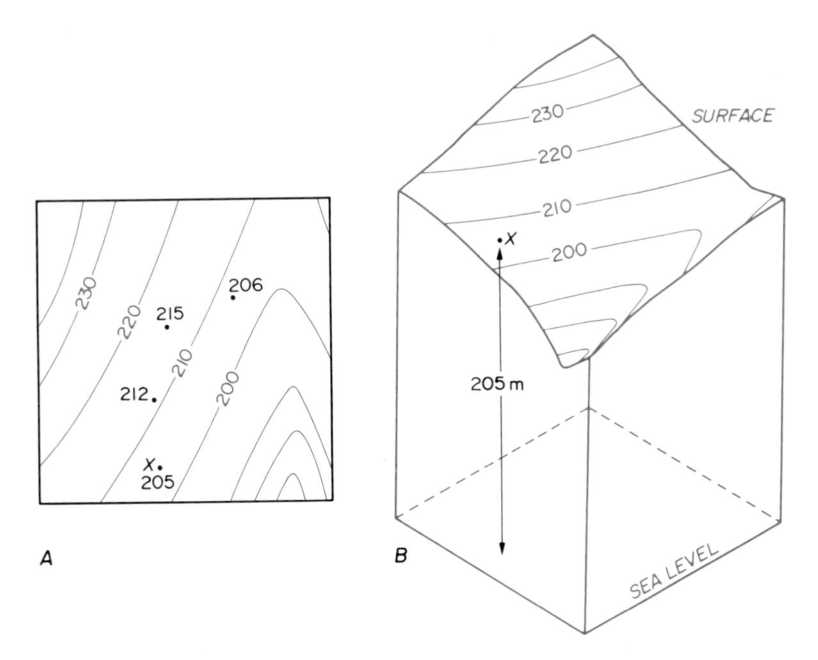

FIG. 2.11 A surface and its representation by means of contours.

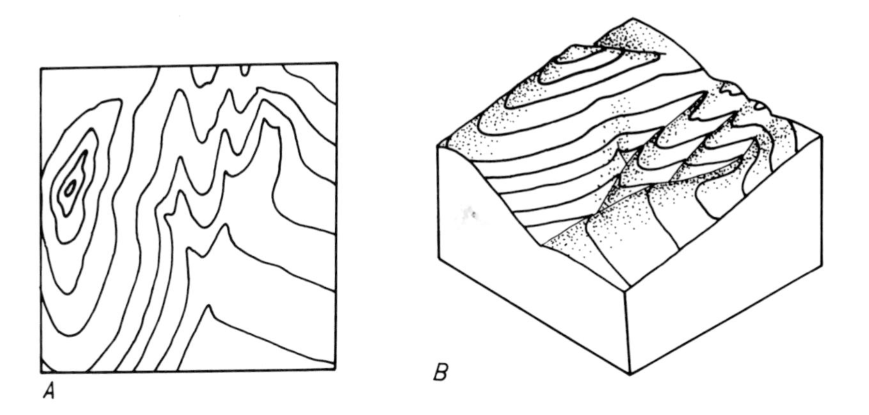

FIG. 2.12 Contour patterns and the form of a surface.

(dipping) surface is represented by parallel equally spaced contours. Isolated hills (dome-shaped structures) will yield closed concentric arrangements of contours and valleys and ridges give V-shaped contour patterns (compare Figs 2.12A and B).

2.9 Drawing vertical cross-sections through topographical and geological surfaces

Vertical *cross-sections* represent the form of the topography and geological structure as seen on a 'cut' through the earth. This vertical cut is imaginary rather than real, so the construction of such a cross-section usually involves a certain amount of interpretation.

The features displayed in the cross-section are the lines of intersection of the section plane with topographical and geological surfaces. When contour patterns are known for these surfaces the drawing of a cross-section is straightforward. If a vertical section is to be constructed between the points X and Y on Fig. 2.13, a base line of length XY is set out. Perpendiculars to the base line at X and Y are then drawn which are graduated in terms of height (Fig. 2.13B). Points on the map where the contour lines for the surface intersect the line of section (line XY) are easily transferred to the section, as shown in Fig. 2.13B.

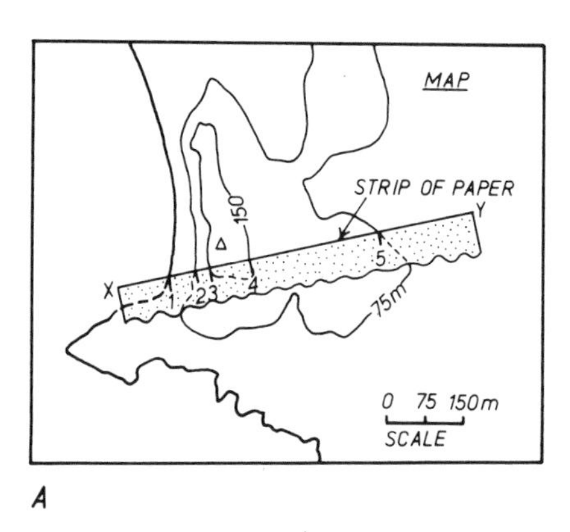

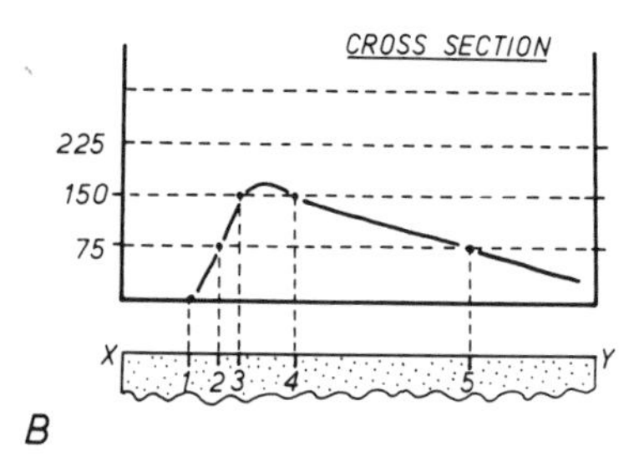

FIG. 2.13 Construction of a cross-section showing surface topography.

Provided the vertical scale used is the same as the horizontal scale, the angle of slope will be the correct slope corresponding to the chosen line of section. For example, if the surface being drawn is geological, the slope in the section will equal the apparent dip appropriate for the line of section. If an exaggerated vertical scale is used, the gradients of lines will be steepened and the structures will also appear distorted in other respects (see Chapter 3 on Folds). The use of exaggerated vertical scales on cross-sections should be avoided.

Worked example

Vertical sections. Figure 2.14A shows a set of structure contours for the surface defined by the base of a sandstone bed. Find the direction of strike, the direction of dip and the angle of dip of the base of the sandstone bed. What is the apparent dip in the direction XZ (Fig. 2.14B)?

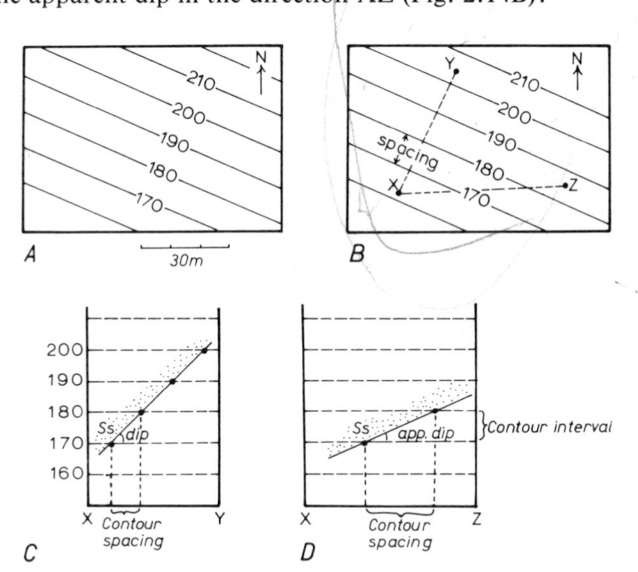

FIG. 2.14 Drawing sections.

The *strike* of the surface at any point is given by the trend of the contours for that surface. On Fig. 2.14A the trend of the contours measured with a protractor is 120°N.

The *dip direction* is 90° away from the strike direction; giving 030° and 210° as the two possible directions of dip. The heights of the structural contours decrease towards the southwest, which tells us that the surface slopes down in that direction. The direction 210° rather than 030° must therefore be the correct dip direction.

To find the *angle of dip* we must calculate the inclination of a line on the surface at right angles to the strike. A constructed vertical cross-section along a line XY on Fig. 2.14B (or any section line parallel to XY) will tell us the true dip of the base of the sandstone. This cross-section (Fig. 2.14C) reveals that the angle of dip is related to the spacing of the contours: i.e.

$$\text{Tangent (angle of dip)} = \frac{\text{contour interval}}{\text{spacing between contours}} = \frac{10\text{m}}{10\text{m}} = 1$$

Therefore, angle of dip = 45°.

The *apparent dip* in direction XZ is the observed inclination of the sandstone bed in true scale (vertical scale = horizontal scale) vertical section along the line XZ. The same formula can be used as for the angle of dip above except 'spacing between contours' is now the apparent spacing we see along the line XZ.

2.10 Three-point problems

Above we have considered a surface described by contours. If, instead of contours, a number of spot heights are given for a surface, then it is possible to infer the form of the contours. This is desirable since surfaces represented by contours are easier to visualize. The number of spot heights required to make a sensible estimate of the form of the contour lines depends on the complexity of the surface. For a surface which is planar, a minimum of three spot heights are required.

Worked example.

A sandstone-shale contact encountered at three localities *A, B* and *C* on Fig. 2.15A has heights of 150, 100 and 175 metres respectively. Assuming that the contact is planar, draw structure contours for the sandstone-shale contact.

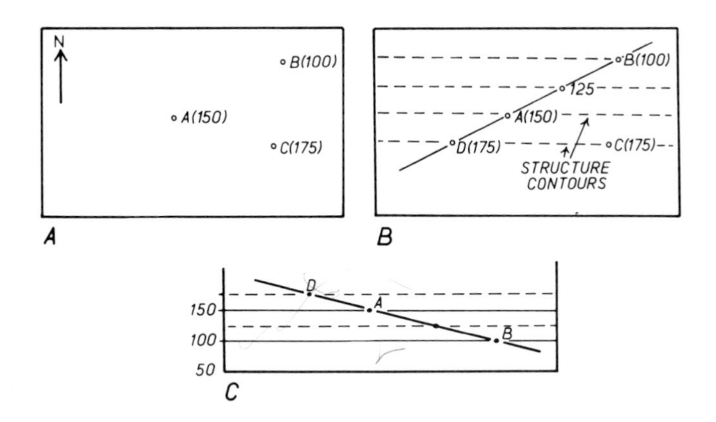

FIG. 2.15 Solution of a three-point problem.

Consider an imaginary vertical section along line *AB* on the map. In that section the contact will appear as a straight line, since it is the line of intersection of two planes: the planar geological contact and the section plane. Furthermore, in that vertical section the line representing the contact will pass through the points *A* and *B* at their respective heights (Fig. 2.15C). The height of the contact decreases at a constant rate as we move from *A* to *B*. This allows us to predict the place along line AB where the surface will have a specified height (Fig. 2.15B). For instance, the contact will have a height of 125 metres at the mid-point between *A* (height equals 150 metres) and *B* (height equals 100 metres). In this way we also locate the point *D* along *AB* which has the same height as the third point *C* (175 metres). In a section along the line *CD* the contact will appear horizontal. Line *CD* is therefore parallel to the horizontal or strike line in the surface. We call *CD* the 175 metre structure contour for the surface. Other structure contours for other heights will be parallel to this, and will be equally spaced on the map. The 100 metre contour must pass through *B* . If it is required to know the dip of the contact the method of the previous example can be used.

2.11 Outcrop patterns of geological surfaces exposed on the ground

We have seen how both the land surface and a geological surface (such as a junction between two formations) can be represented by contour maps. The line on a geological map representing the contact of two formations marks the intersection of these two surfaces. The form of this line on the map can be predicted if the contour patterns defining the topography and the geological surface are known, since along the line of intersection both surfaces will have equal height.

Worked example

Given topographic contours and structure contours for a planar coal seam (Fig. 2.16A) predict the map outcrop pattern of the coal seam.

Points are sought on the map where structure contours intersect a topographic contour of the same elevation. A series of points are obtained in this way through which the line of outcrop of the coal seam must pass. This final stage of joining the points to form a surface outcrop would seem in places to be somewhat arbitrary with the lines labelled *p* and *q* in Fig. 2.16B being equally possible. However *p* is incorrect, since the line of outcrop cannot cross the 700 metre structure contour unless there is a point along it at which the ground surface has a height of 700 metres.

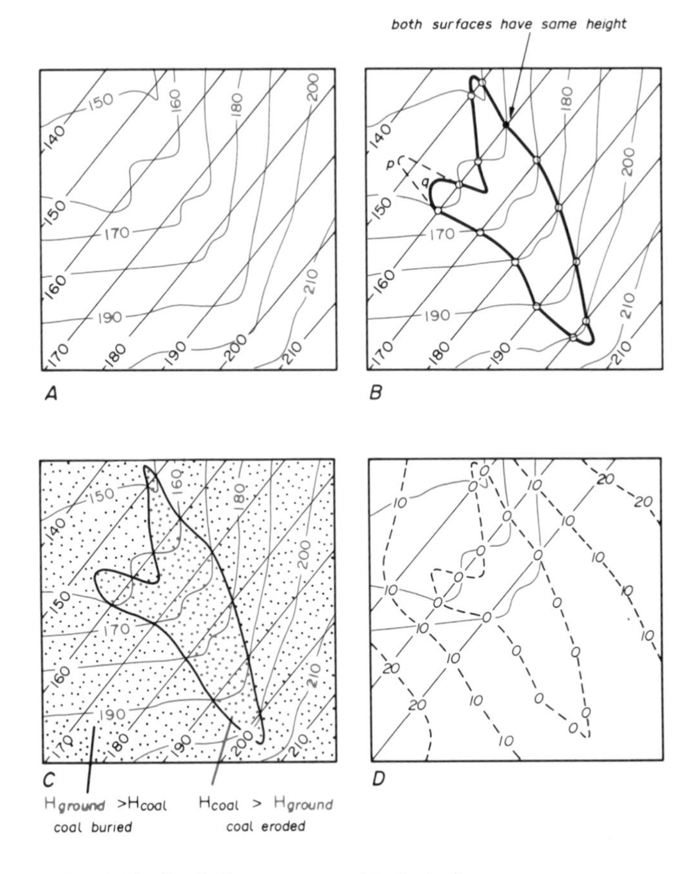

FIG. 2.16 Predicting outcrop and isobaths from structure contour information.

2.12 Buried and eroded parts of a geological surface

The coal seam in the previous example only occurs at the ground surface along a single line. The surface at other points (a point not on the line of outcrop) is either buried (beneath ground level) or eroded (above ground level). The line of outcrop in Fig. 2.16B divides the map into two kinds of areas:

(a) areas where height (coal) > height (topography), so that the surface can be thought to have existed above the present topography but has been eroded away, and

(b) areas where height (coal) < height (topography) so that the coal must exist below the topography, i.e. it is buried.

The boundary line between these two types of areas is given by the line of outcrop, i.e. where height (coal) = height (topography).

Worked Example

Using the data on Fig. 2.16A shade the part of the area underlain by coal. The answer to this is shown in Fig. 2.16C. The outcrop line of the coal forms the boundary of the area underlain by coal. The sought area is where the contours for the topography show higher values than the contours of the coal.

2.13 Contours of burial depth (isobaths)

A geological surface is buried below the topographic surface when height (topography) > height (geological surface). The difference (height topography minus height geological surface) equals the depth of burial at any point on the map. Depths of burial determined at a number of points on a map provide data which can be contoured to yield lines of equal depth of burial called *isobaths*.

Worked example

Using again the data from Fig. 2.16A construct isobaths for the coal seam. In the area of buried coal, determine spot depths of coal by subtracting the height of the coal seam from the height of the topography at a number of points. Isobaths, lines linking all points of equal depth of burial, can then be drawn (dashed lines in Fig. 2.16D).

Devise an alternative way to draw isobaths, noting that a 100 metre isobath for a given geological surface is the line of outcrop on an imaginary surface which is everywhere 100 metres lower than the ground surface.

2.14 V-shaped outcrop patterns

A dipping surface which crops out in a valley or on a ridge will give rise to a V-shaped outcrop (Fig. 2.17). The way the outcrop patterns vee depends on the dip of the geological surface relative to the topography. In the case of valleys, patterns vee upstream or downstream (Fig. 2.17). The rule for determining the dip from the type of vee (the 'V rule') is easily remembered if one considers the intermediate case (Fig. 2.17D) where the outcrop vees in neither direction. This is the situation where the dip is equal to the gradient of the valley bottom. As soon as we tilt the beds away from this critical position they will start to exhibit a V-shape. If we visualize the bed to be rotated slightly upstream it will start to vee upstream, at first veeing more sharply than the topographic contours defining the valley (Fig. 2.17C). The bed can be tilted still further upstream until it becomes horizontal. Horizontal beds always yield outcrop patterns which parallel the topographic contours and hence, the beds still vee upstream (Fig. 2.17B). If the bed is tilted further again upstream, the beds starts to dip upstream and we retain a V-shaped

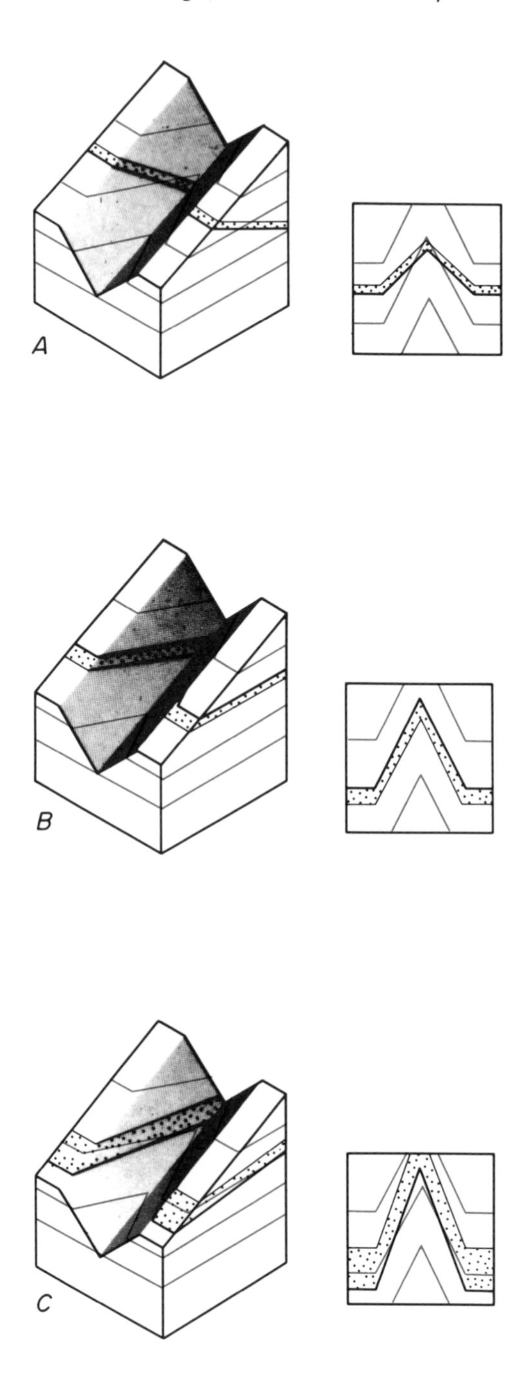

FIG. 2.17 To illustrate the V-rule.

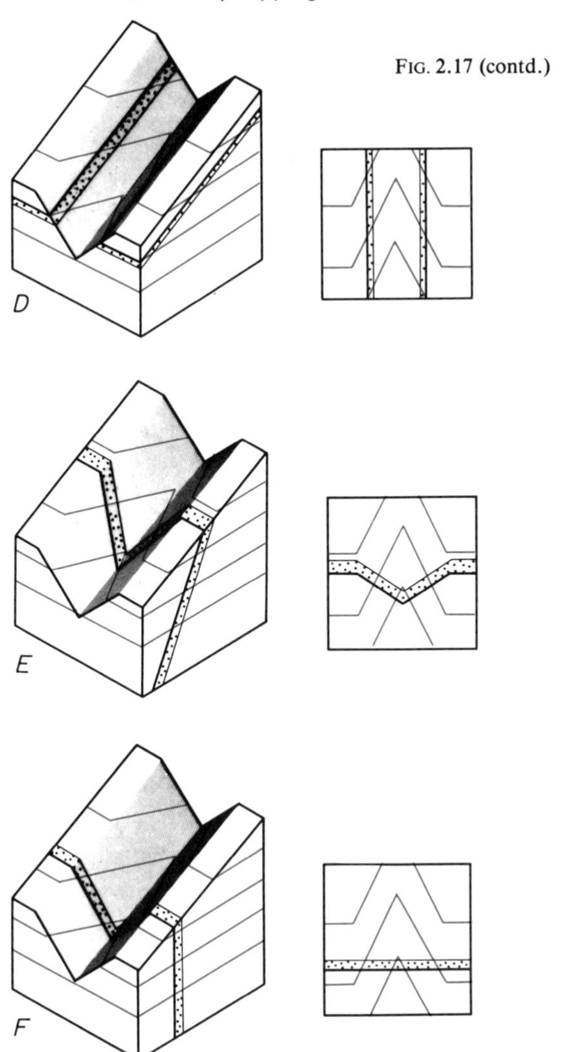

FIG. 2.17 (contd.)

outcrop but now the vee is more 'blunt' than the vee exhibited by the topographic contours (Fig. 2.17A).

Downstream-pointing vees are produced when the beds dip downstream more steeply than the valley gradient (Fig. 2.17E). Finally, vertical beds have straight outcrop courses and do not vee (Fig. 2.17F).

Worked example

Complete the outcrop of the thin limestone bed exposed in the northwest part of the area (Fig. 2.18A). The dip of the bed is 10° towards 220° (220/10).

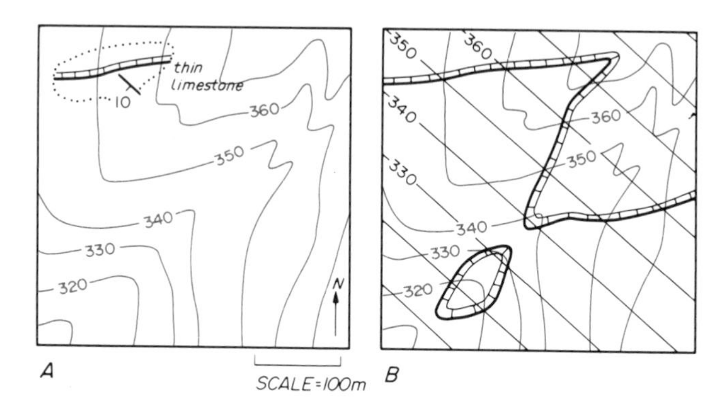

FIG. 2.18 Interpreting the shape of a geological contact in an area of limited rock exposure.

This type of problem is frequently encountered by geological mappers. On published geological maps all contacts are shown. However, rocks are not everywhere exposed. Whilst mapping, a few outcrops are found at which contacts are visible and where dips can be measured, but the rest of the map is based on interpretation. The following technique can be used to interpret the map. Using the known dip, construct structure contours for the thin bed. These will run parallel to the measured strike and, for a contour interval of 10 metres, will have a spacing given by this equation (see Section 2.9).

$$\text{Tangent (angle of dip)} = \frac{\text{contour interval}}{\text{spacing between contours.}}$$

Since the outcrop of the bed in the northwest part of the map is at a height of 350 metres, the 350 metre structure contour must pass through this point. Others are drawn parallel at the calculated spacing. The crossing points of the topographic contours with the structure contours of the same height, yield points which lie on the outcrop of the thin limestone bed. The completed outcrop of the thin limestone bed is shown in Figure 2.18B.

2.15 Structure contours from outcrop patterns

A map showing outcrops of a surface together with topographic contours can be used to construct structure contours for that surface. The underlying principles are:

(a) where a surface outcrops, height surface = height topography,
(b) if the height of a planar surface is known at a minimum of three places, the structure contours for that surface can be constructed (see 'Three-point problems', Section 2.10).

Worked example

Draw strike lines for the limestone bed in Fig. 2.19A. What assumptions are involved?

Join points on the outcrop which share the same height. These join lines are structure contours for that particular height (Fig. 2.19B).

Draw as many structure contours as possible to test the assumption of constant dip (planarity of surface).

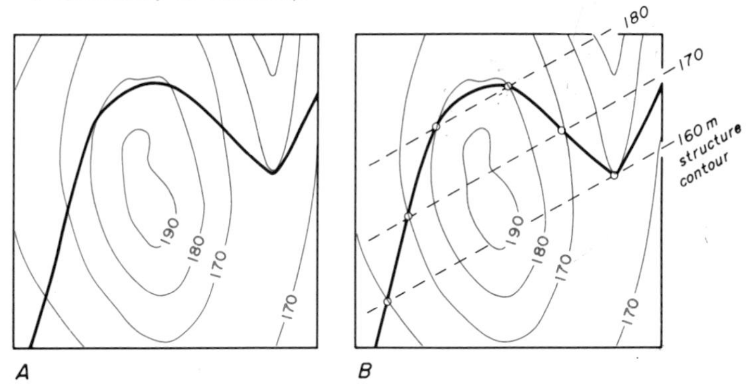

FIG. 2.19 Drawing structure contours.

2.16 Geological surfaces and layers

So far in this chapter the geological structure considered has consisted of a single surface such as the contact surface between two rock units. However formations of rock, together with the individual beds of sediments from which they are composed, are tabular in form and have a definite thickness. Such 'layers' can be dealt with by considering the two bounding surfaces which form the contacts with adjacent units.

2.17 Stratigraphic thickness

The *true or stratigraphic thickness* of a unit is the distance between its bounding surfaces measured in a direction perpendicular to these surfaces (TT in Fig. 2.20).

The *vertical thickness* (*VT*) is more readily calculated from structure contour maps. The vertical thickness is the height difference between the top and base of the unit at any point. It is the vertical 'drilled' thickness, and is obtained by subtracting the height of the base from the height of the top. Depending on the angle of dip (α), the vertical thickness (*VT*) differs from the true thickness (*TT*), because from Fig. 2.20B:

$$\cos \alpha = TT/VT$$

and $$TT = VT \cos \alpha.$$

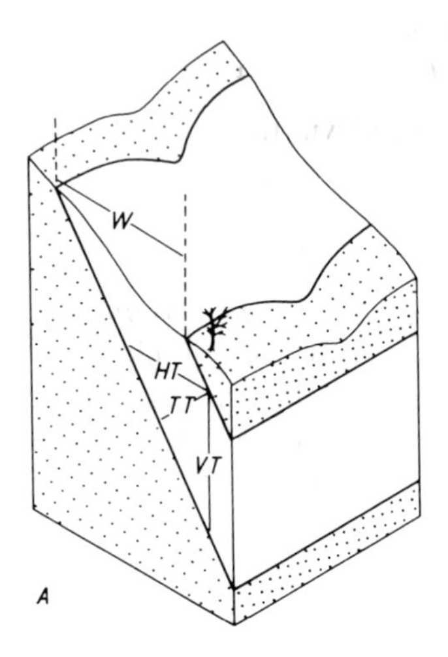

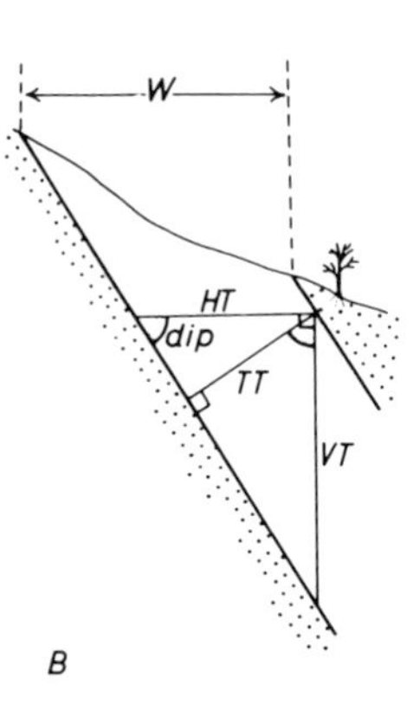

FIG. 2.20 Bed thickness and width of outcrop.

This equation can be used to calculate the true thickness if the vertical thickness is known. The *horizontal thickness* (HT) is a distance measured at right angles to the strike between a point on the base of the unit and a point of the same height on the top of the unit. It can be found by taking the separation on the map between a contour line for the base of the bed and one for the top of the bed of the same altitude.

From Fig. 2.20B it can be also seen that

$$\sin \alpha = TT/HT$$

therefore,

$$TT = \sin \alpha \, HT.$$

If VT and HT are both known, the dip can be calculated from

$$\tan \alpha = \frac{VT}{HT}.$$

It is important to note that the *width of outcrop* of a bed on a map (W on Fig. 2.20) is not equal to the horizontal thickness unless the ground surface is horizontal. In cross-sections care must be taken when interpreting thicknesses. Vertical thickness will be correct in any vertical section but the true thickness will only be visible in cross-sections parallel to the dip direction of the beds.

Worked example

Find the vertical thickness, horizontal thickness, true thickness and angle of dip of the sandstone formation from the structure contours in Fig. 2.21.

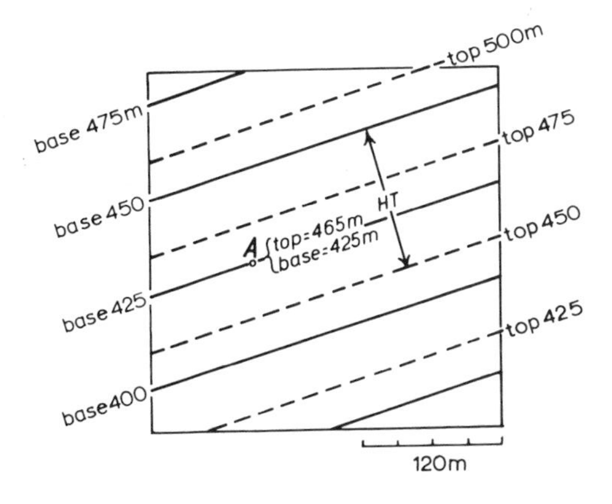

FIG. 2.21 Calculation of thickness from structure contours.

The vertical thickness is obtained by taking any point on the map, say A, and using the equation:

Vertical thickness = height of top − height of base = 465 − 425 = 40 m. The horizontal thickness is given by horizontal separation of any pair of structure contours of the same altitude (one for base, one for top). The horizontal thickness in this example is 120 m.

2.18 Isochores and isopachs

Contour lines and isobaths are examples of lines drawn on a map which join points where some physical quantity has equal value. *Isochores* are lines of equal vertical thickness and *isopachs* are lines of equal stratigraphic (true) thickness.

2.19 Topographic effects and map scale

If the surface of the earth's surface were everywhere horizontal, geological map reading would be much easier, since all contacts on the map would run parallel to their strikes. For a geological surface it is the existence of slopes in the landscape which causes the discrepancy between its course on the map and the direction of strike. This 'terrain effect' is most marked on a smaller scale because natural ground slopes are generally steeper at this scale. On 1:10,000 scale maps for instance, the presence of valleys and ridges exercises a strong influence on the shape of all but the steepest dipping surfaces. On the other hand the run of geological boundaries on, say, the 1:625,000 geological map of the United Kingdom, is a direct portrayal of the local strike of the rocks. Only where dips are gentle and relief is high (e.g. the Jurassic outcrops of the Cotswolds) does the 'terrain effect' play any significant role. The generally lower average slopes of the earth's surface at this scale makes the interpretation of the map pattern much more straightforward.

PROBLEM 2.1

The photograph (Problem 2.1) shows dipping Carboniferous Limestone beds at Brandy Cove, Gower, South Wales. The north direction is shown by an arrow in the sand.

(a) What is the approximate direction of the strike of these beds? (Give a compass direction.)
(b) What is the approximate angle of dip?
(c) Write down the attitude of the bedding as a single expression of the form:
Dip direction/angle of dip

PROBLEM 2.2

The map shows outcrops on a horizontal topographic surface.

Interpret the run of the geological boundaries and complete the map.

Draw the structure on the three vertical faces of the block diagram (below).

Label the following on the completed block diagram:

(a) angle of dip
(b) angle of apparent dip
(c) the strike of the beds
(d) the direction of dip of the beds

PROBLEM 2.3

An underground passage linking two cave systems follows the line of intersection of the base of a limestone bed and a vertical rock fracture. The bedding in the limestone dips 060/60 and the strike of the fracture is 010°. What is the inclination (plunge) of the underground passage?

PROBLEM 2.4

An imaginary London to Swansea railway has a number of vertical cuttings which run in an east-west direction.

At Port Talbot, Coal Measures rocks dip 010/30; near Newport, Old Red Sandstone rocks dip 315/20; and at Swindon, Upper Jurassic rocks have the dip 160/10.

At which cutting will railway passengers observe the steepest dip of strata (apparent dips are observed in the cuttings)?

PROBLEM 2.5

The map shows structure contours for the basal contact of a mudstone bed.

What is the strike of the contact?

What is the dip direction of the contact?

What is the angle of dip of the contact?

Construct an east-west true scale cross-section (equal vertical and horizontal scales) to show the contact.

Explain why the angle of dip seen in the drawn section differs from the dip calculated above.

Use a formula to calculate the dip in the section and to check the accuracy of the cross-section.

PROBLEM 2.6

For each map, determine the direction and angle of dip of the geological contact shown.

PROBLEM 2.7

This is a geological map of part of the Cotswolds. Examine the relationship between geological boundaries and topographic contours and deduce the dip of the rocks.

Deduce as much as possible about the thicknesses of the Jurassic formations exposed in this area.

Draw a cross-section along the line *A-B*.

PROBLEM 2.8

The topographic map shows an area near Port Talbot in West Glamorgan. In three boreholes drilled in Margam Park the 'Two-Feet-Nine' coal seam was encountered at the following elevations above sea level:

Borehole	Height above sea level
Margam Park No. 1	– 110 m
Margam Park No. 2	– 150 m
Margam Park No. 3	– 475 m

Draw structure contours for the Two-Feet-Nine seam, assuming it maintains a constant dip within the area covered by the map.

What is the direction of dip and angle of dip of the Two-Feet-Nine seam?

Does the seam crop out within the area of the map?

A second seam (the 'Field Vein') occurs 625 m above the Two-Feet-Nine seam. Construct the line of outcrop of the Field Vein.

PROBLEM 2.9

The map shows a number of outcrops where a sandstone/mudstone contact has been encountered in the field.

How do the data available support the notion that the contact has a uniform dip?

Interpret the run of the contact through the rest of the area covered by the map.

Calculate:

a) the direction of dip expressed as a compass bearing and

b) the angle of dip

PROBLEM 2.10

The base of the Lower Greensand is encountered in three boreholes in Suffolk at the following heights above sea level:

−150 m at the Culford Borehole (Map Ref. 831711)
−75 m at the Kentford Borehole (Map Ref. 702684)
−60 m at the Worlington Borehole (Map Ref. 699738)

Construct structure contours for the base of the Lower Greensand.

Predict the height of the base of the Lower Greensand below the Cathedral at Bury St Edmunds (856650).

Where, closest to Bury St Edmunds, would the base of the Lower Greensand be expected to crop out if the topography in the area of outcrop is more or less flat at a height of 50 m above sea level?

If a new borehole at Barrow (755635) were to encounter Lower Greensand at height −100 m, how would it affect your earlier conclusions.

PROBLEM 2.11

For each map predict the outcrop of a thin bed which occurs at *A*. In each map the bed has a different dip. On map (a) the dip is 11° northwards, on map (b) the dip is 11° southwards, in (c) the dip is vertical and the strike is east-west, and in (d) the dip is 3.4° southwards. (Note: $11° = \tan^{-1}(0.2)$ and $3.4° = \tan^{-1}(0.06)$.)

PROBLEM 2.12

Dipping Jurassic strata, southeast of Rich, Morocco

(a) Draw a geological sketch map based on the photograph (kindly provided by Professor M. R. House). On this map show the general form of topographic contours together with the outcrop pattern of a number of the exposed beds. (Note the way the dipping strata 'vee' over the ridge in the foreground.)

(b) Re-write the V-rule in Section 2.14 of this chapter to express the way the outcrop pattern of beds exposed *on a ridge* varies depending on the dip of the strata.

PROBLEM 2.13

Draw a cross-section of the map between points *A* and *B*.

Calculate the 30, 60, 90 and 120 m isobaths for the 'Rhondda Rider' coal seam.

Calculate the stratigraphic and vertical thickness of the Llynfi Beds.

PROBLEM 2.1

PROBLEM 2.2

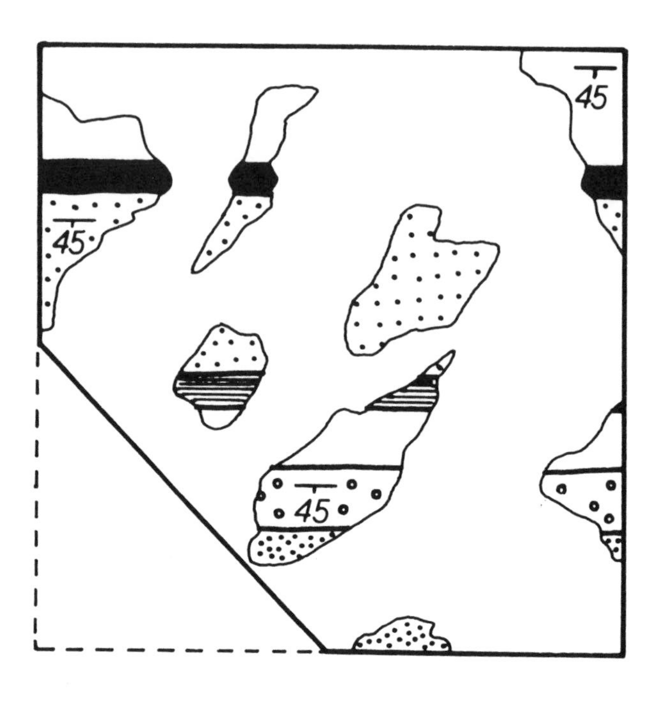

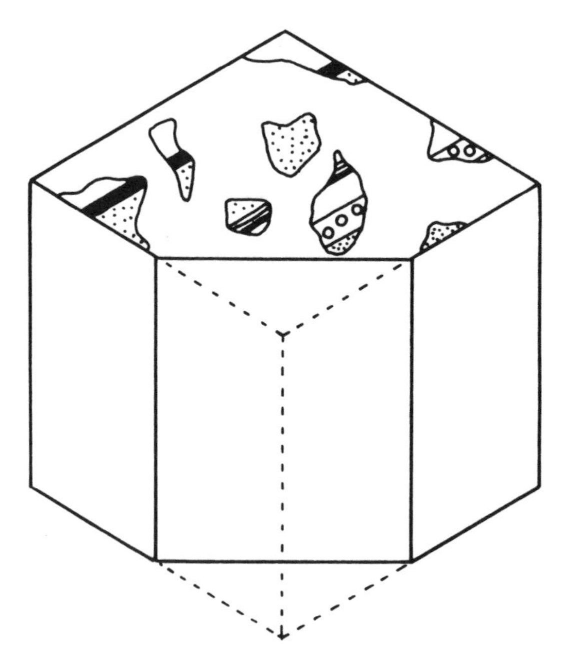

PROBLEM 2.5

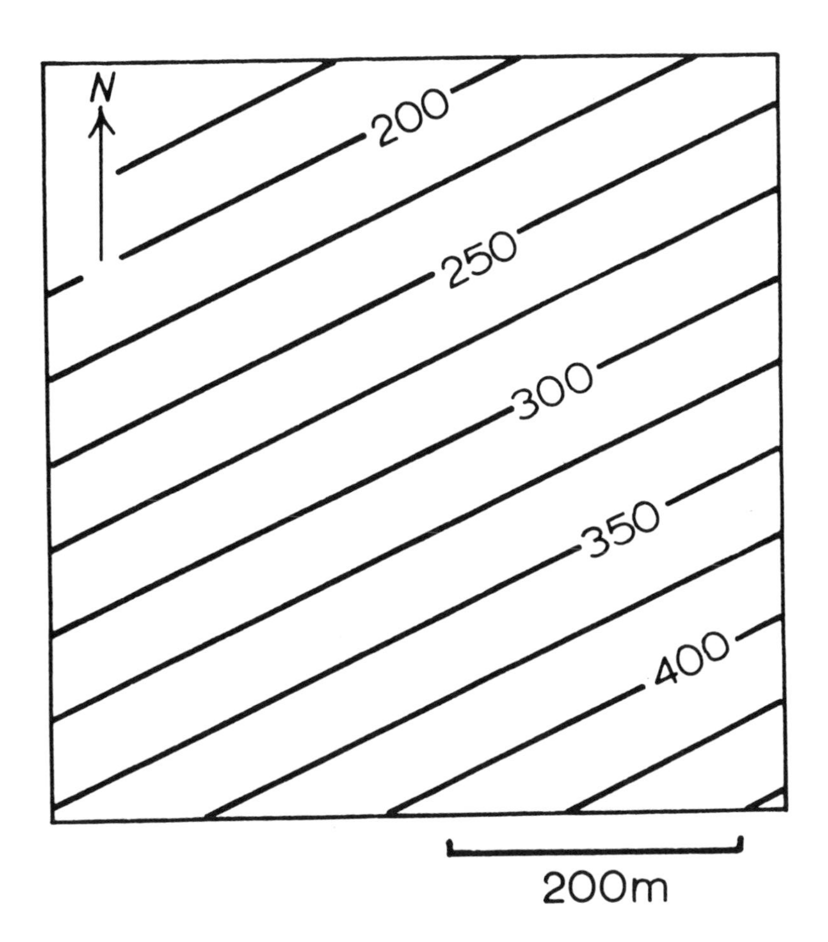

PROBLEM 2.6

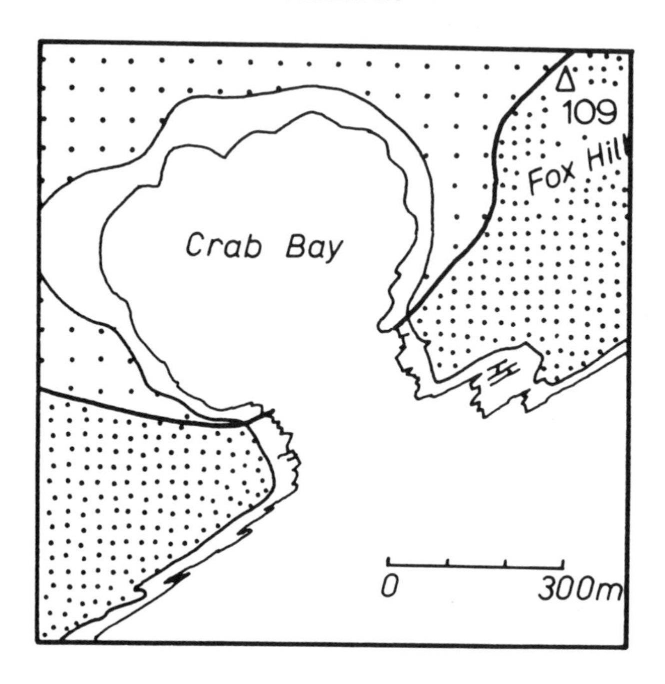

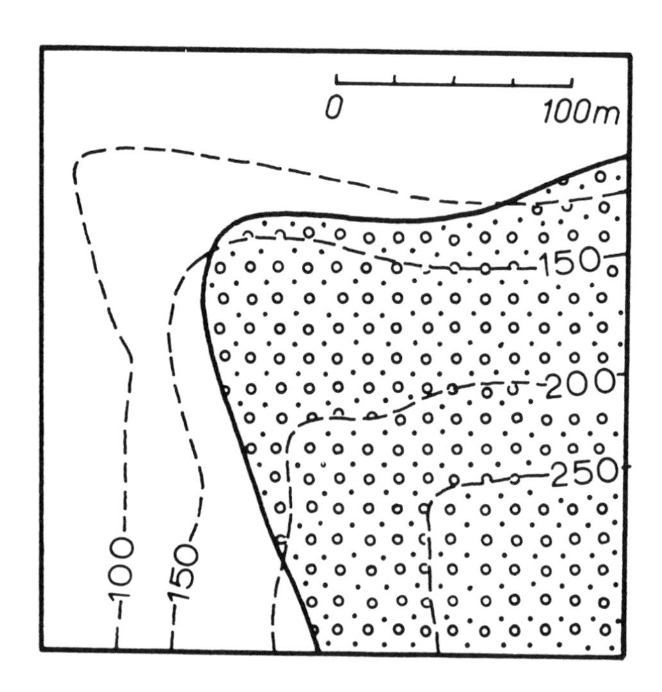

PROBLEM 2.7

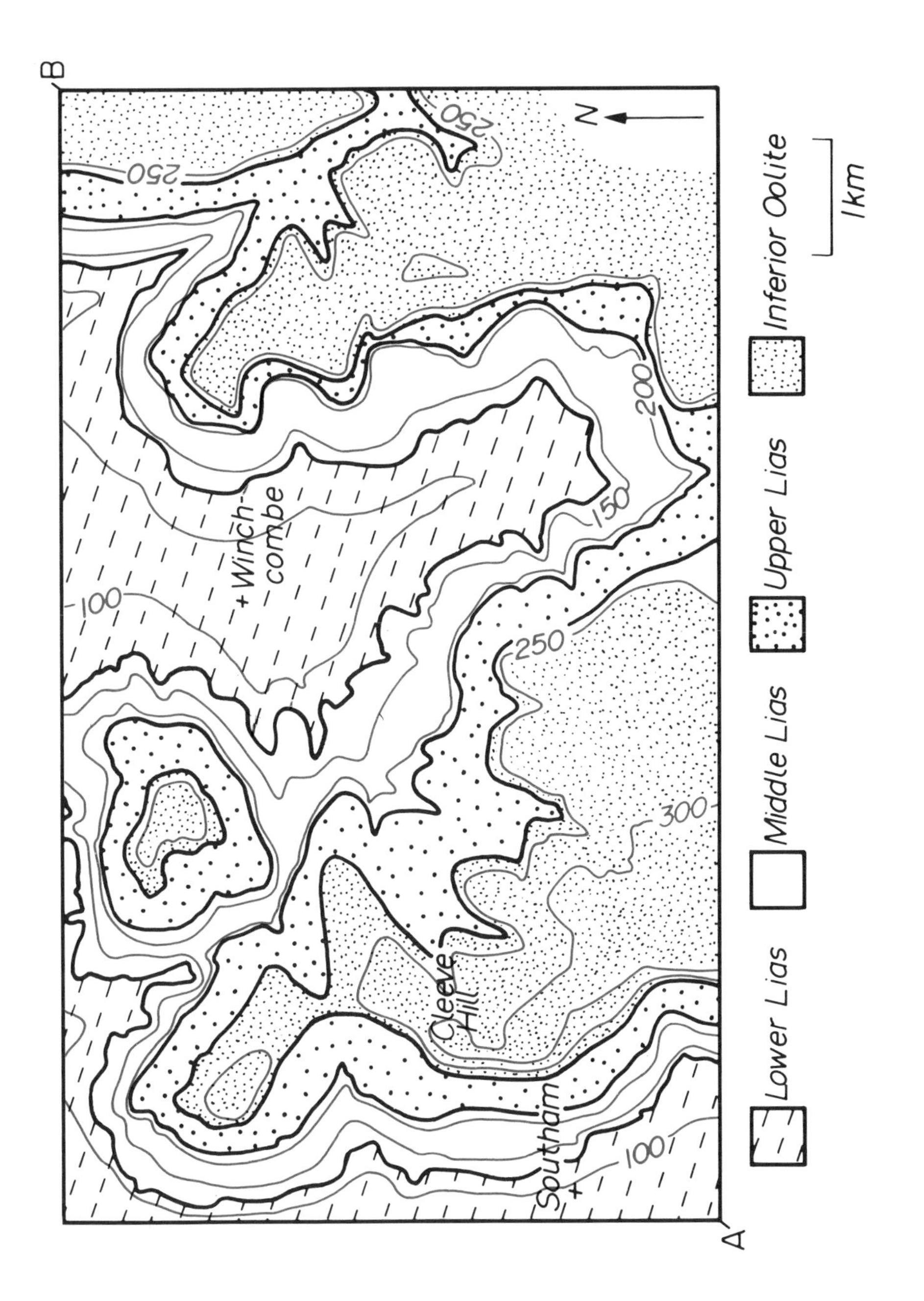
B
A
N
250
250
200
150
100
250
300
100
Winch-combe
Cleeve Hill
Southam
Lower Lias
Middle Lias
Upper Lias
Inferior Oolite
1km

PROBLEM 2.8

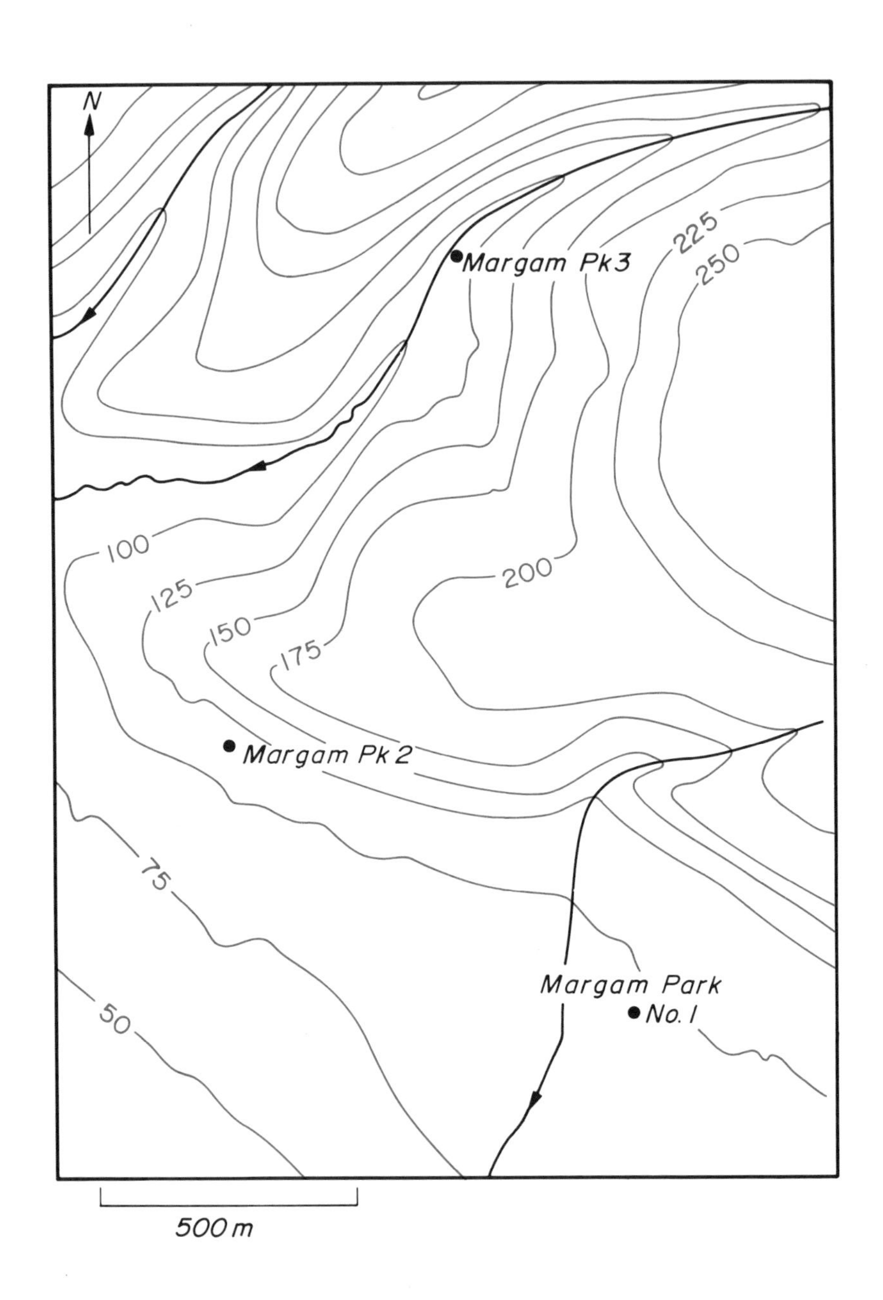
N
Margam Pk3
225
250
100
200
125
150
175
Margam Pk 2
75
Margam Park
No. 1
50
500 m

PROBLEM 2.9

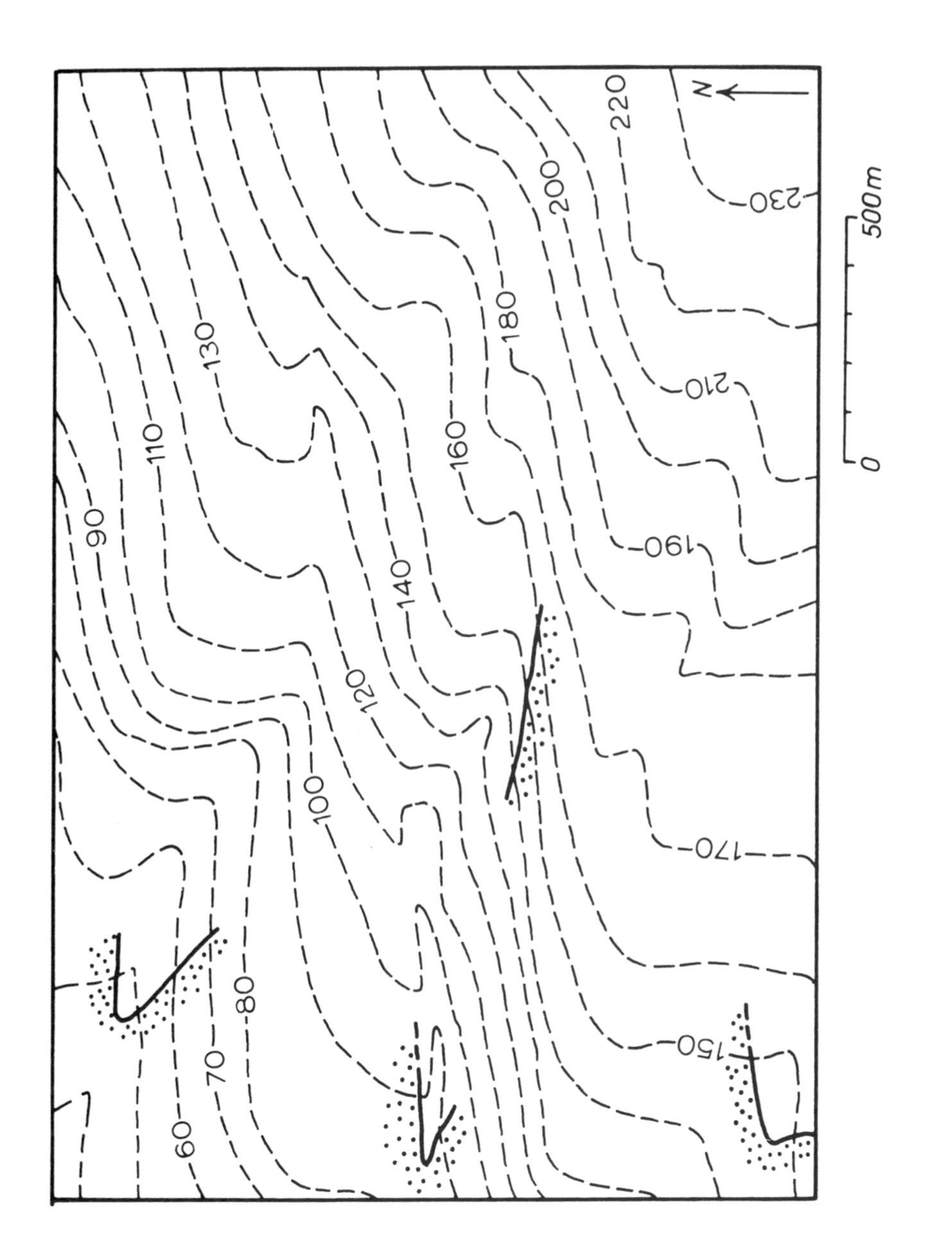
N
0
500m
230
220
210
200
190
180
170
160
150
140
130
120
110
100
90
80
70
60

PROBLEM 2.10

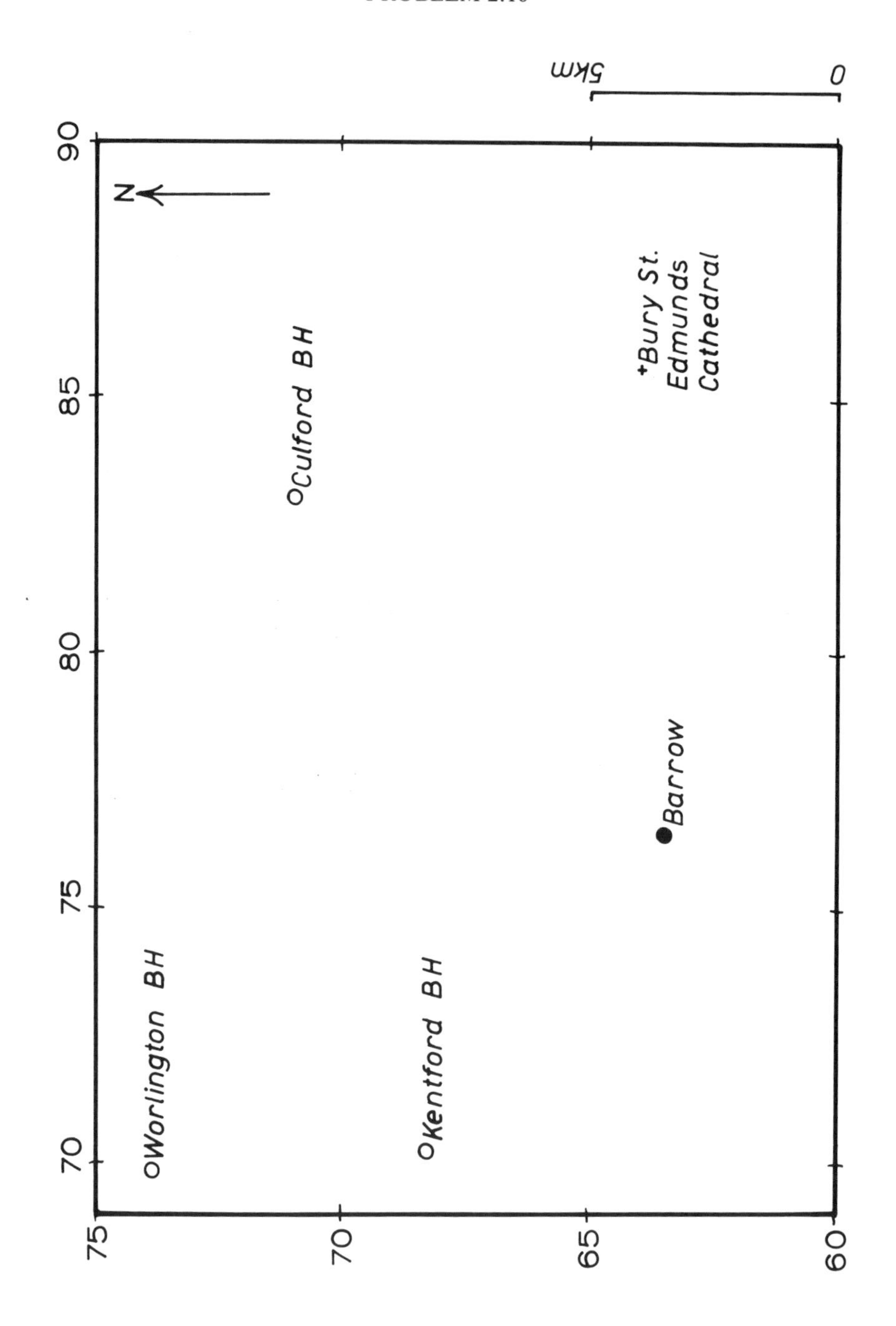
0
5km
N
90
85
80
75
70
75
70
65
60
Culford BH
Bury St. Edmunds Cathedral
Barrow
Worlington BH
Kentford BH

PROBLEM 2.11

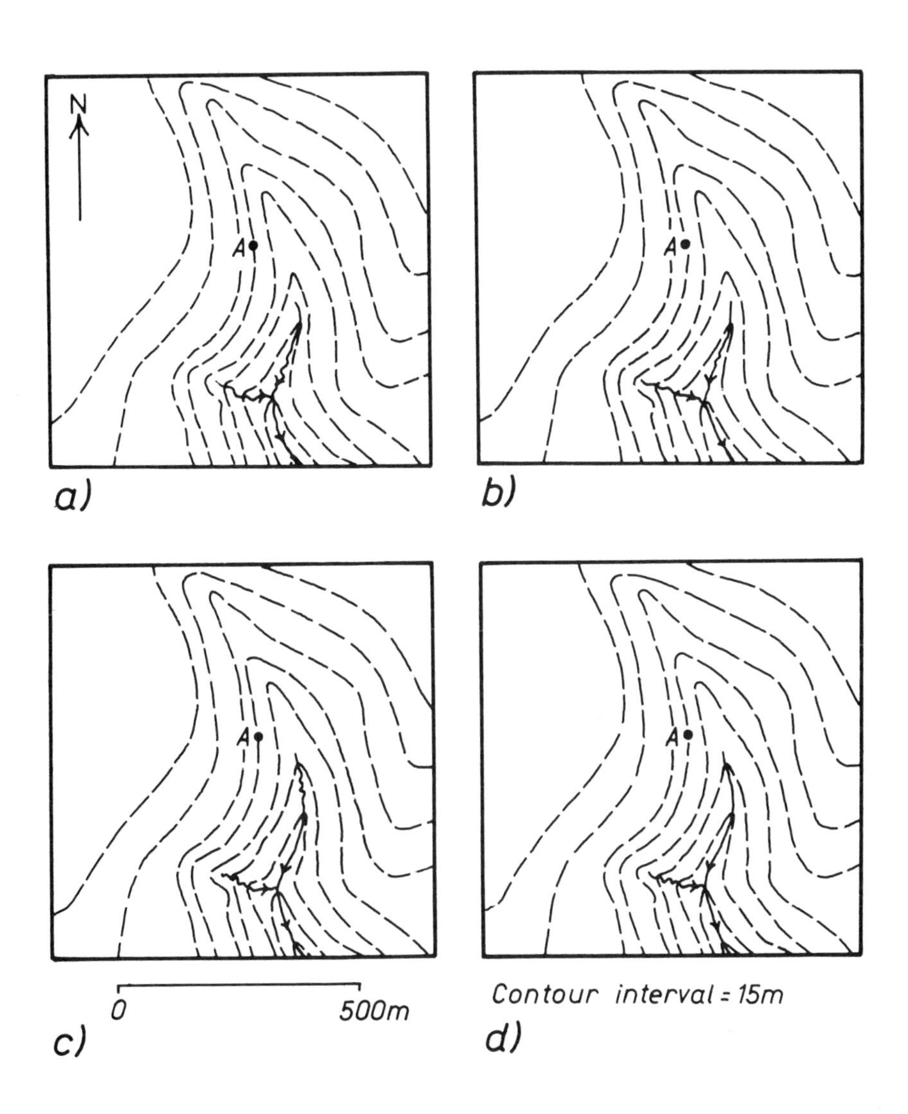

PROBLEM 2.12

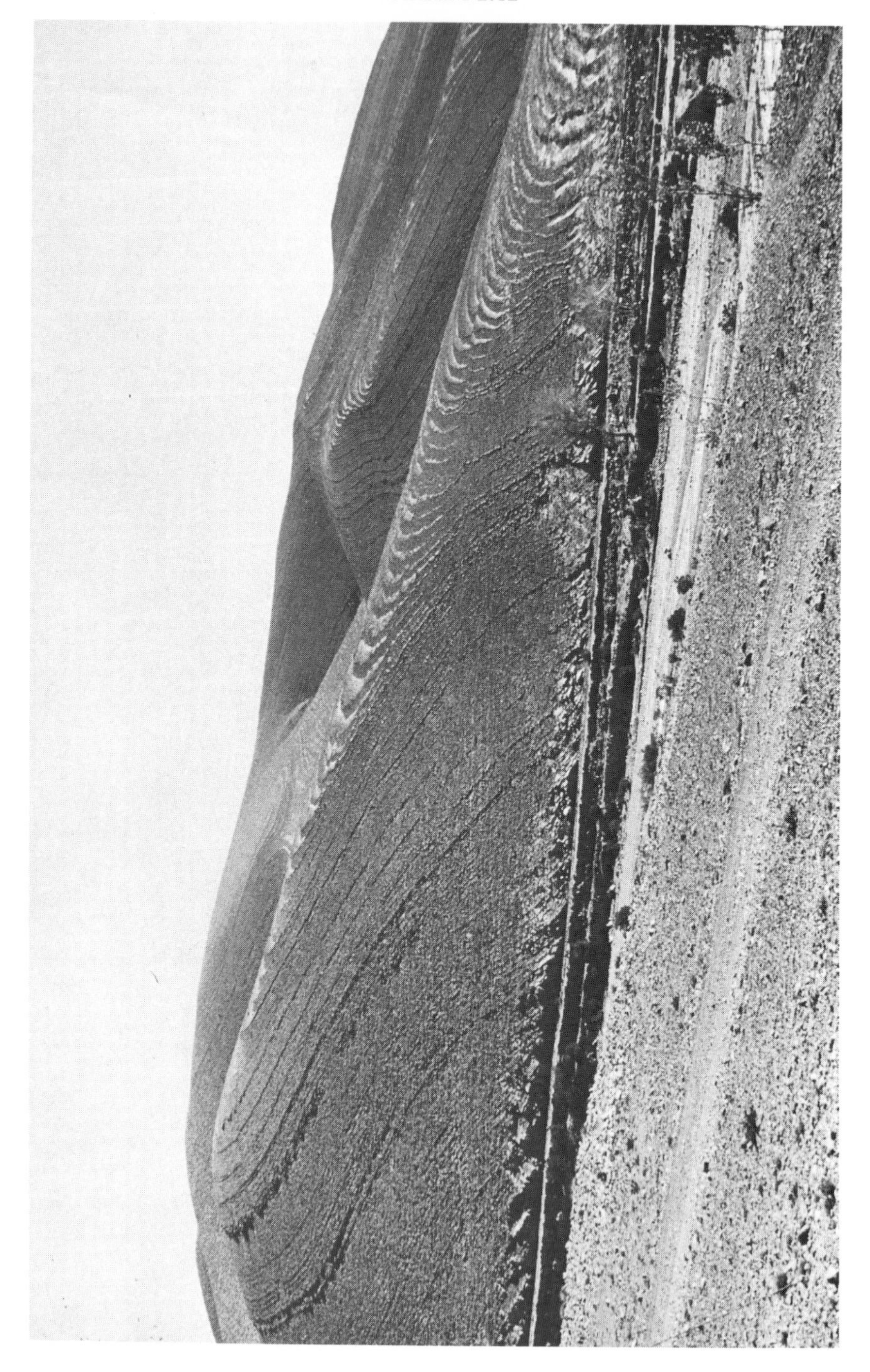

PROBLEM 2.13

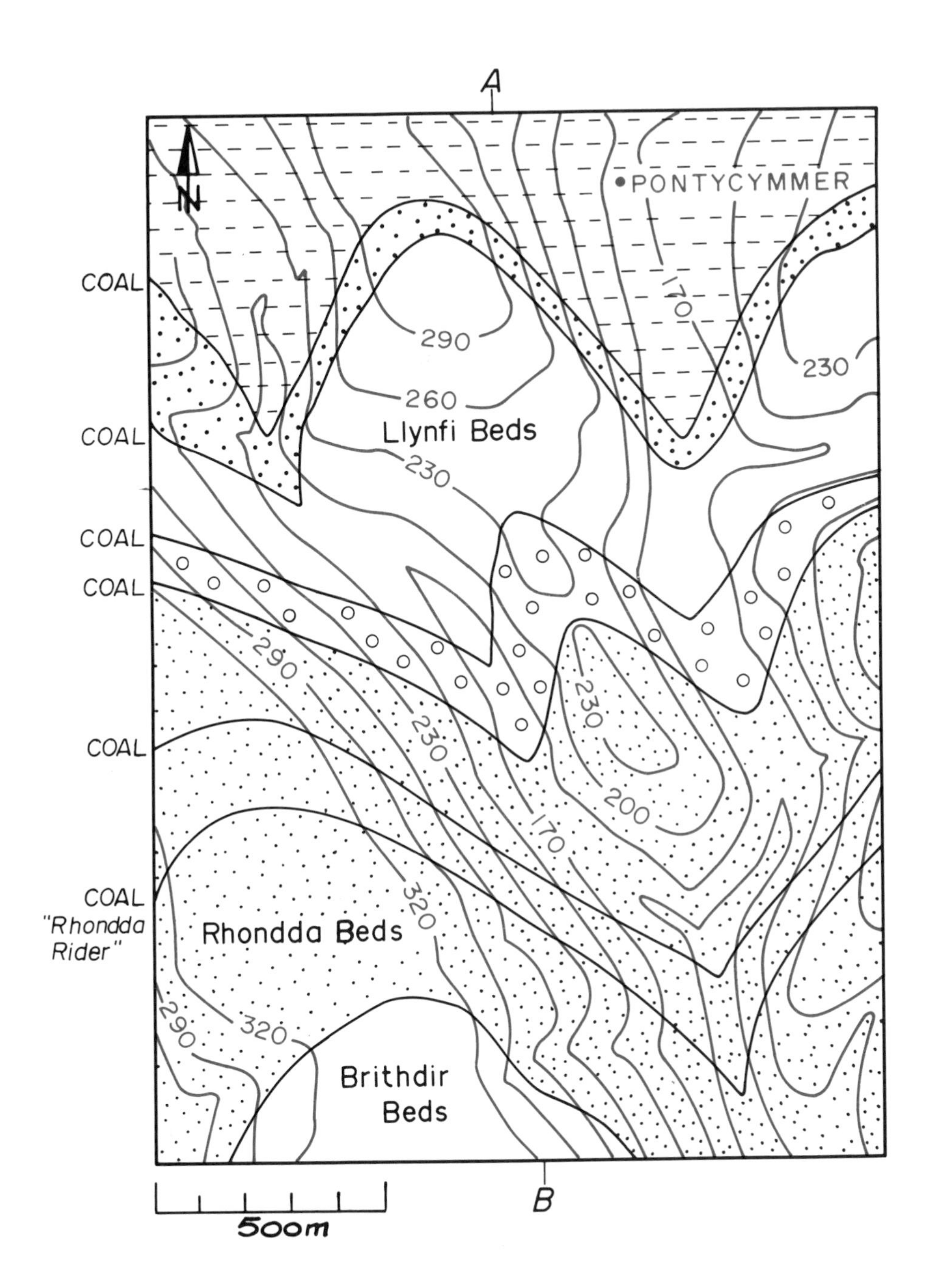
A
N
PONTYCYMMER
COAL
COAL
COAL
COAL
COAL
COAL
"Rhondda Rider"
170
290
260
230
Llynfi Beds
230
290
230
230
200
170
320
Rhondda Beds
290
320
Brithdir Beds
B
500m

3

Folding

THE previous chapter dealt with planar geological surfaces. A geological surface which is curved is said to be *folded.* Most folding is the result of crustal deformation whereby rock layering such as bedding has been subjected to a shortening in a direction within the layering. To demonstrate this place both hands on a tablecloth and draw them together; the shortening of the tablecloth results in a number of folds.

The structure known as folding is not everywhere developed equally. For example the Upper Carboniferous rocks of southwestern England and southernmost Wales show intense folding when compared to rocks of the same age further north in Britain. The crustal deformation responsible for the production of this folding was clearly more severe in certain areas. Zones of concentrated deformation and folding are called fold belts or mountain belts, and these occupy long parallel-sided tracts of the earth's crust. For example, the present-day mountain chain of the Andes is a fold belt produced by the shortening of the rocks of South America since the end of the Cretaceous times.

3.1 Cylindrical and non-cylindrical folding

A curved surface, the shape of which can be generated by taking a straight line and moving it whilst keeping it parallel to itself in space, is called a *cylindrically folded surface* (Fig. 3.1). A corrugated iron roofing sheet or a row of greenhouse rooves have the form of a set of cylindrical folds. Folds which cannot be generated by translating a straight line are called *non-cylindrical.* An example of this type of shape is an egg-tray. Figures 3.2 and 3.4 show examples of cylindrical and non-cylindrical folds. The line capable of 'generating' the surface of a cylindrical fold is called the *fold axis.*

An important property of cylindrical folds is that the fold shape, as viewed on serial sections (cross-section planes which are parallel like those made by a ham slicer), remains constant (Fig. 3.3A). This is true whatever the attitude of the section plane. Serial sectioning of a non-cylindrical fold produces two-dimensional fold shapes which vary from one section to another (Fig. 3.3B).

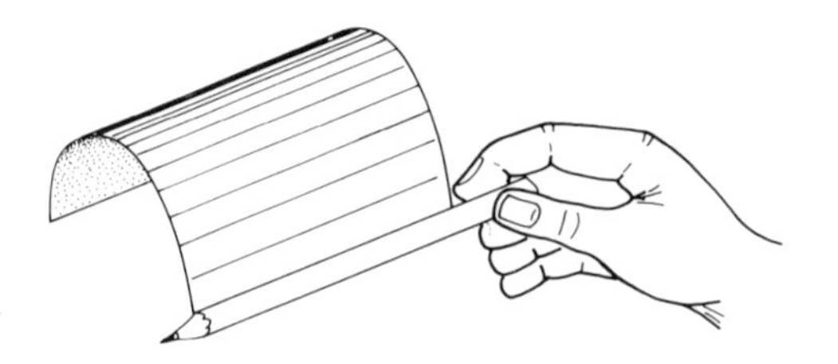

FIG. 3.1 The concept of a cylindrically folded surface.

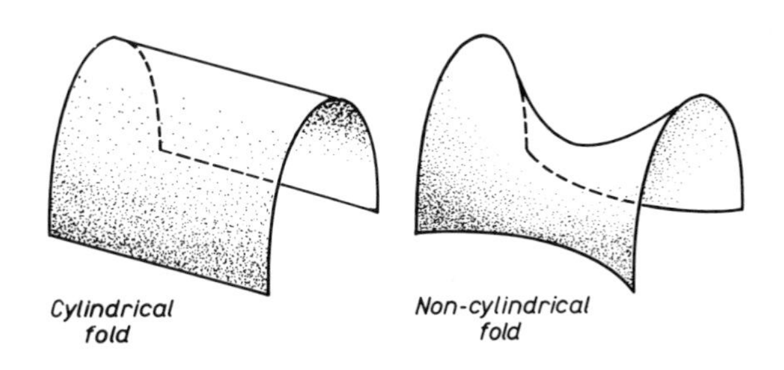

FIG. 3.2 **A**: Cylindrical fold. **B**: Non-cylindrical fold.

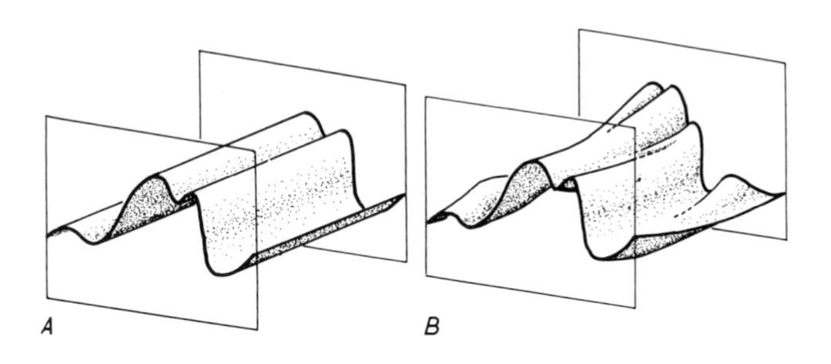

FIG. 3.3 Parallel sections through (**A**) cylindrical folds, (**B**) non-cylindrical folds.

FIG. 3.4 **A**: Cylindrical folding.

FIG. 3.4 **B**: Non-cylindrical folding.

3.2 Basic geometrical features of a fold

The single curved surface in Fig. 3.5 shows three folds. The lines which separate adjacent folds are the *inflection lines*. They mark the places where the surface changes from being convex to concave or vice-versa. Between adjacent inflection lines the surface is not uniformly convex or concave but there are places where the curvature is more pronounced. This is called the *hinge zone*. The *hinge line* is the line of maximum curvature. Like the inflection lines, the hinge line need not be straight except when the folding is cylindrical. Hinge lines divide folds into separate *limbs*.

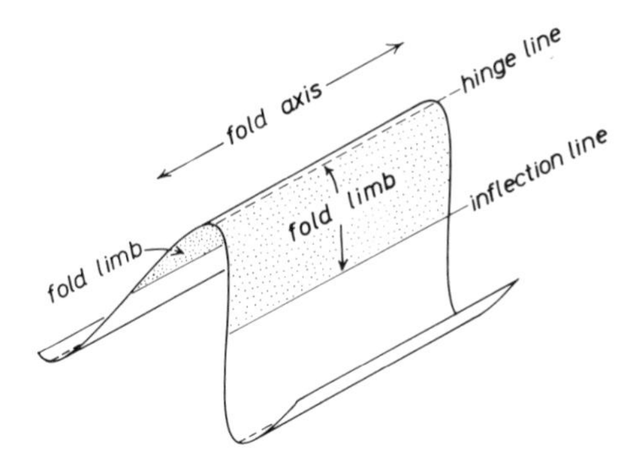

FIG. 3.5 Some folding terms.

The terms introduced so far can be used for a single surface such as a single folded bedding plane. Folding usually affects a layered sequence so that a number of surfaces are folded together. *Harmonic folding* is where the number and positions of folds in successive surfaces broadly match (Figs. 3.6A and 3.7B). Where this matching of folds does not exist the style of folding is called *disharmonic* (Fig. 3.6B).

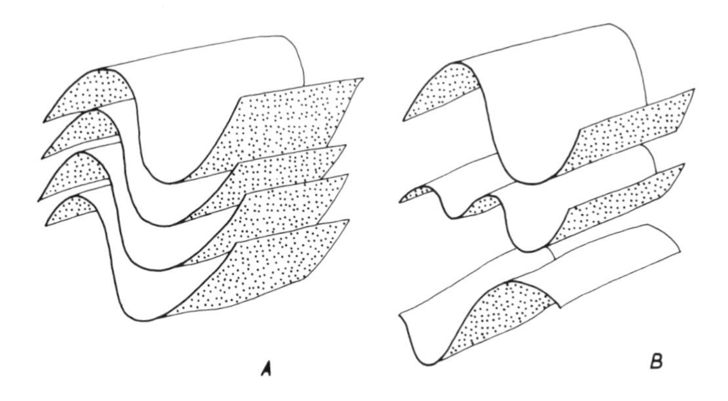

FIG. 3.6 **A**: Harmonic folding. **B**: Disharmonic folding.

FIG. 3.7 Folds in the field. **A**: Upright folding with hinge lines which plunge towards the camera (Precambrian metasediments, Anglesey).

FIG. 3.7 **B**: Upright harmonic folding (Nordland, Norway).

The *axial surface* of a fold is the surface which contains the hinge lines of successive harmonically folded surfaces (Fig. 3.8). For obvious reasons this surface is sometimes referred to as the hinge surface. The axial surface need not be planar but is often curved.

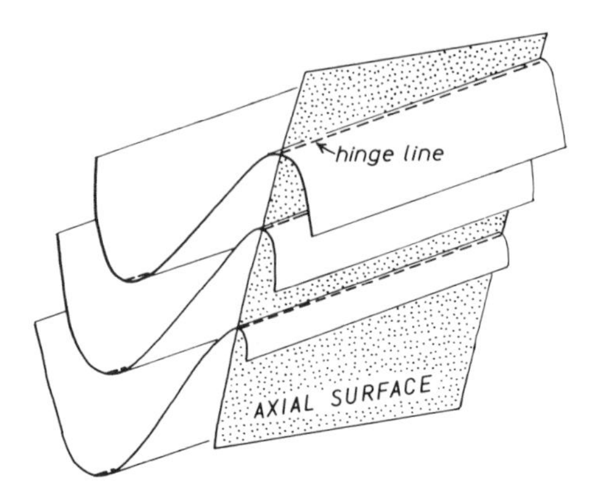

FIG. 3.8 The axial surface.

3.3 Terms relating to the orientation of folds

As a result of folding, the overall length of a bed as measured in a direction perpendicular to the axial surface is shortened. When folding in a particular region is being studied great attention is paid to the direction of folds, since this is indicative of the direction in which the strata have been most shortened.

The orientations of the fold hinge and the axial surface are the two most important directional characteristics of a fold. The fold hinge, which in the case of a cylindrical fold is parallel to the fold axis, is a linear feature. As explained in Section 2.3, such linear features have an orientation which is described by the plunge direction and the angle of plunge.

Non-plunging folds have horizontal hinges or plunges of less than 10° whilst *vertical folds* plunge at 80-90° (Fig. 3.9). The orientation of the axial surface is described by means of its dip direction and angle of dip. When the axial plane dips less than 10° the adjective *recumbent* is applicable. Folds are *inclined* when the axial plane dips and the term *upright* is reserved for folds with steeply dipping (>80°) axial surfaces (Fig. 3.9). Figure 3.7A shows a field example of upright plunging folds.

In any fold the orientations of the hinge line and axial surface are not independent attributes. Because the fold hinge, by definition, is a direction which lies within the axial surface, its maximum possible plunge is limited by the angle of dip of the axial surface. For a given axial surface the steepest possible hinge-line plunge is obtained when the fold plunges in the direction of dip of the axial surface. Such folds are called *reclined* (Fig. 3.9).

Another directional feature of the fold is the direction in which the limbs of the fold converge or close. This *direction of closure* is a direction within the axial surface at right angles to the fold hinge (Fig. 3.10).

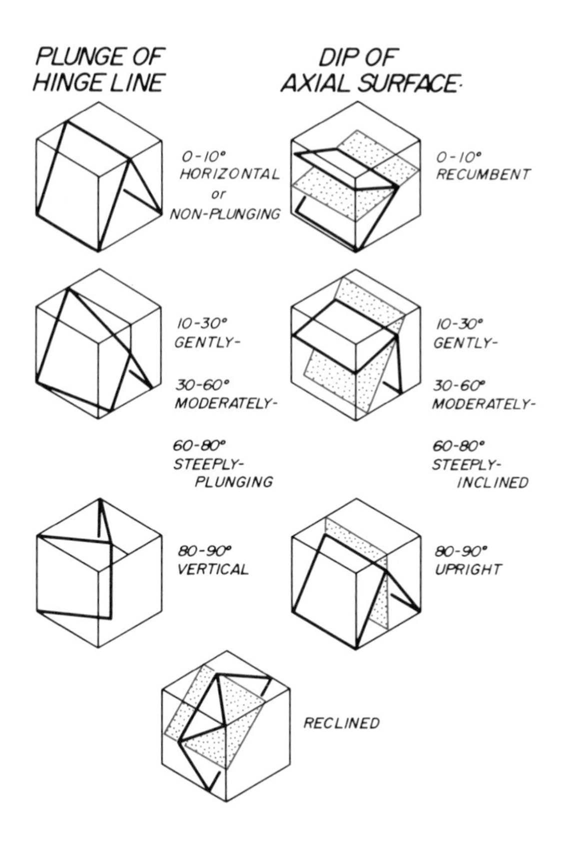

FIG. 3.9 Names given to folds depending on their orientation.

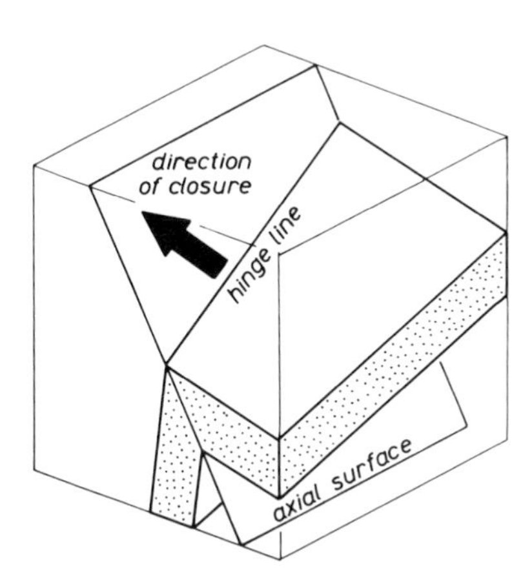

FIG. 3.10 Direction of closure.

On the basis of the direction of closure, three fold types are distinguished:

antiforms: close upwards
synforms: close downwards
neutral folds: close in a horizontal direction.

Exercise

Which of the folds in Fig. 3.9 are neutral folds? (The answer is given at the end of this chapter.)

Where folding affects a sequence of beds, and where their ages are known, the facing of the folds can be determined. The *direction of facing* is the direction in the axial surface at right angles to the fold hinge line pointing towards the younger beds (Fig. 3.11). Folds may face upwards, downwards or sideways.

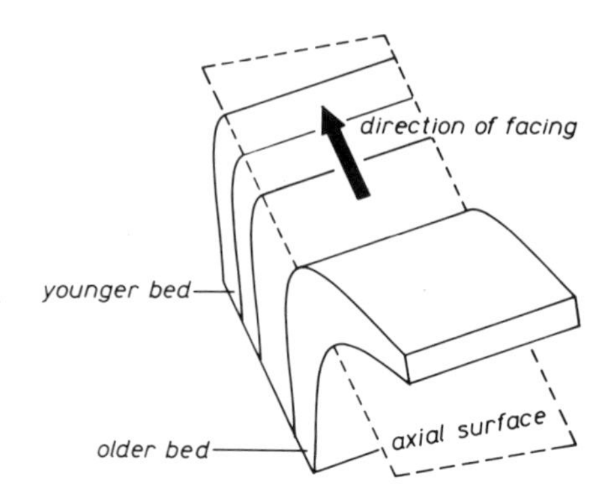

Fig. 3.11 How the direction of facing is defined.

In *anticlines* the facing direction is towards the outer arcs of the fold, i.e. away from the core of the fold (Fig. 3.12). *Synclines* face towards their inner arcs or cores. Anticlines can be either antiformal, synformal or neutral. The same is true of synclines.

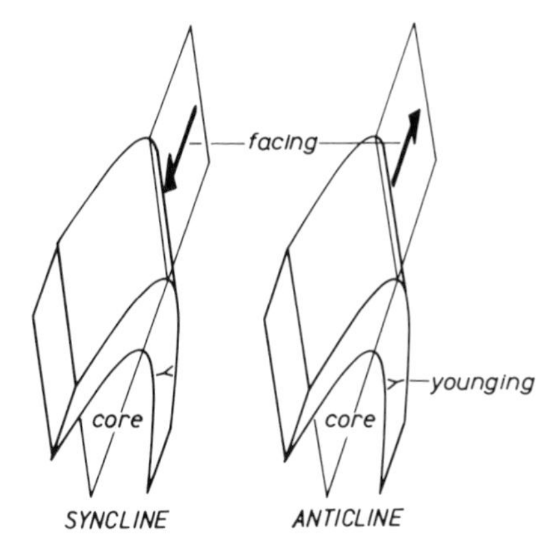

FIG. 3.12 In a syncline the beds young towards the folds core; in an anticline the beds young away from the core of the fold. Either type can be antiformal or synformal.

3.4 The tightness of folding

Folds can be classified according to their degree of openness/tightness. The *tightness* of a fold is measured by the size of the angle between the fold limbs. The *interlimb angle* is defined as the angle between the planes tangential to the folded surface at the inflection lines (Fig. 3.13).

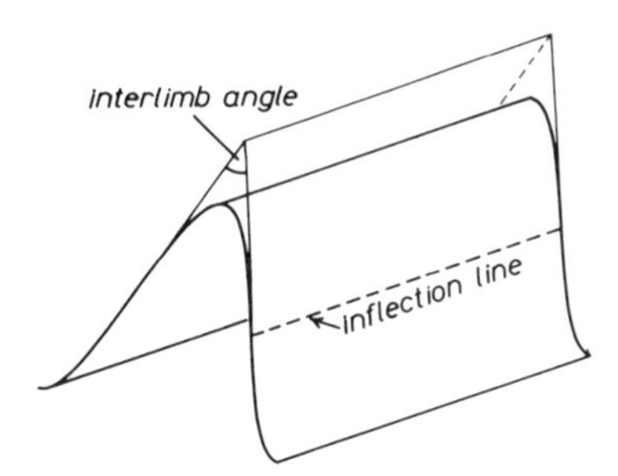

FIG. 3.13 The tightness of fold is determined from the interlimb angle.

The size of the interlimb angle allows the fold to be classified in the following scheme:

Interlimb angle	Description of fold
180° - 120°	Gentle
120° - 70°	Open
70° - 30°	Close
30° - 0°	Tight
0°	Isoclinal
Negative angle	Mushroom

Depending on the dip of the axial surface, tight folds may have limbs which dip in the same general direction. Such folds are called *overturned* folds.

3.5 Curvature variation around the fold

The three folds in Fig. 3.14 have the same tightness since they possess the same interlimb angle. Nevertheless the shapes of the fold differ significantly due to a different distribution of curvature around the fold. Figure 3.14A shows a fold with a fairly constant curvature. This rounded shape contrasts with angular fold in Fig. 3.14C. Chevron, accordion, and concertina are all names used for folds of this latter angular type (Fig. 3.16A).

FIG. 3.14 Folds with the same tightness but different hinge curvature.

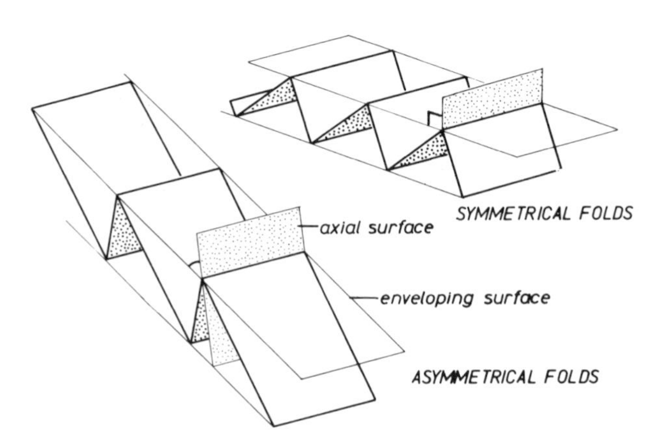

FIG. 3.15 Symmetrical and asymmetrical folds.

FIG. 3.16 **A**: Recumbent chevron folds at Millook, north coast of Cornwall. **B**: Periclinal folds exposed on a horizontal outcrop surface, Precambrian gneisses, Nordland, Norway. The closed oval arrangements of layers also characterizes the map pattern of larger-scale periclinal folds.

3.6 Symmetrical and asymmetrical folds

When one limb of a fold is the mirror image of the other, and the axial surface is a plane of symmetry, the fold is said to be *symmetrical* (Fig. 3.15).

There exists a common misconception that the dips of limbs of a symmetrical fold must have equal dips in opposite directions. This need not be the case, but the lengths of the limbs must be equal (Fig. 3.15). Another property of a symmetrical fold is that the *enveloping surface* (the surface describing the average dip of the folded bed) is at right angles to the axial surface of each fold.

Asymmetrical folds usually have limbs of unequal length and an enveloping surface which is not perpendicular to the axial surface (Fig. 3.15). Figures 3.7 and 3.19A show asymmetrical folds.

3.7 Types of non-cylindrical fold

Domes and *basins* (Fig. 3.17) are folds of non-cylindrical type since their shape cannot be described by the simple translation of a straight line. More commonly occurring non-cylindrical folds possess a well-defined but curved hinge line. Four types having the shapes of whalebacks, saddles, canoes and shoe horns (Fig. 3.17) are called *periclinal folds.* They are doubly-plunging with points on their hinge line called plunge *culminations* and *depressions* where the direction of plunge reverses. Periclinal folding gives rise to closed elliptical patterns on the map or outcrop surface (Fig. 3.16B).

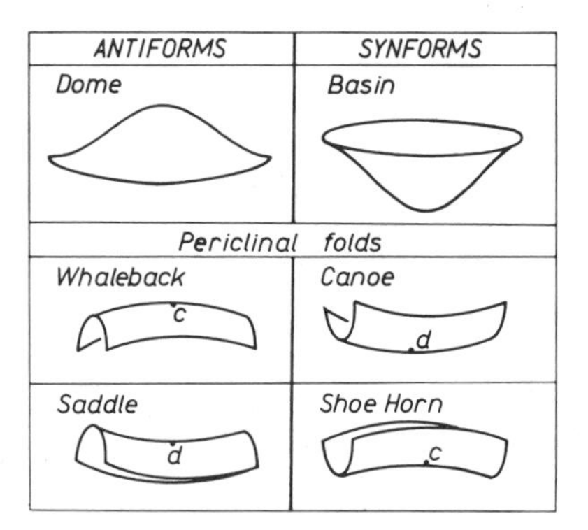

FIG. 3.17 Types of non-cylindrical folds.

3.8 Layer thickness variation around folds

The process leading to the formation of folds from originally planar layers or beds involves more than a simple rotation of the limbs about the hinge line. The layering around the limbs also undergoes distortions or strains which leads to a relative thinning of the layering in some positions in the fold relative to others. The careful measurement of the bed thickness at a number of points between the inflection points provides data which allow the fold to

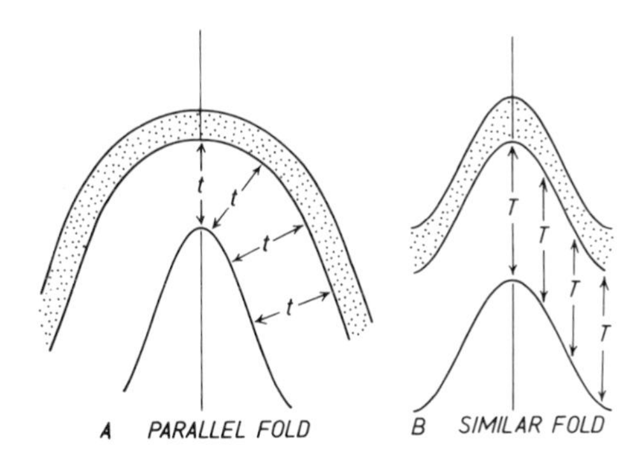

FIG. 3.18 Parallel and similar folds are distinguished by the way the thickness varies around the fold.

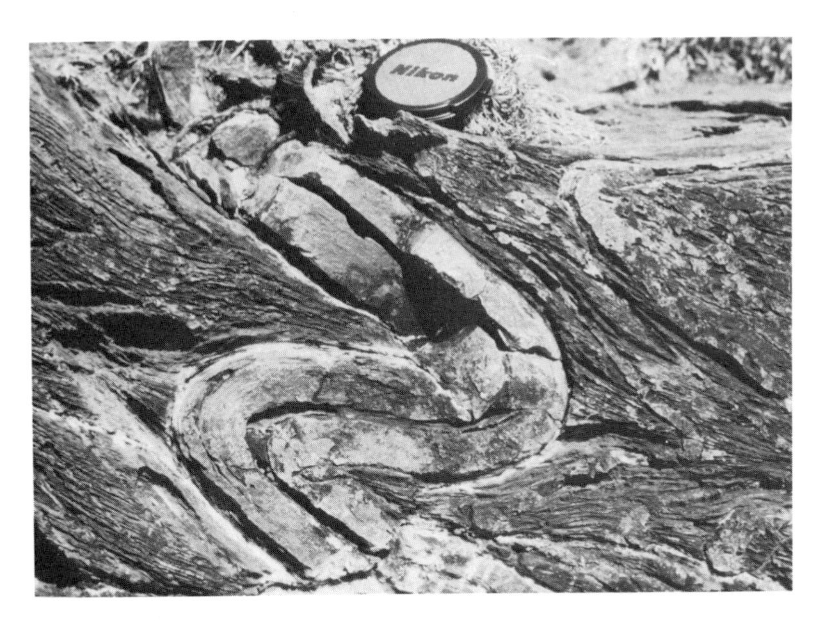

FIG. 3.19 Folds which classify as parallel-type (A) and similar-type (B). Compare the variation of stratigraphic thickness of the layers in each example.

FIG. 3.19 **B**

be classified. One important class of folds in this scheme has constant bed (stratigraphic) thickness and are called *parallel folds* (Figs. 3.18A and 3.19A). In a second class of folds the stratigraphic thickness is greater at the hinge than on the limbs, though the bed thickness is constant if measured in a direction parallel to the axial surface (Figs. 3.18B and 3.19B). The latter are called *similar folds* because the upper and lower surfaces of the bed have identical shapes.

3.9 Structure contours and folds

In the previous chapter it was stated that uniformly dipping surfaces are represented by structure contour patterns consisting of parallel, evenly spaced contour lines. Since the strike and angle of dip of a surface usually vary around a fold, the structure contours are generally curved and variably spaced (Fig. 3.20).

The shape of a set of structure contour lines depicts the shape of horizontal serial sections through the folded surface. Since cylindrical folds give identical cross-sections on parallel sections (see Section 3.1), these folds give structure contour patterns consisting of contours of similar shape and size (Fig. 3.20A). Non-cylindrical folds give rise to more complex structure contour patterns (Fig. 3.20B). Concentric circular contours indicate domes or basins. The various types of periclinal folds have characteristic contour arrangements

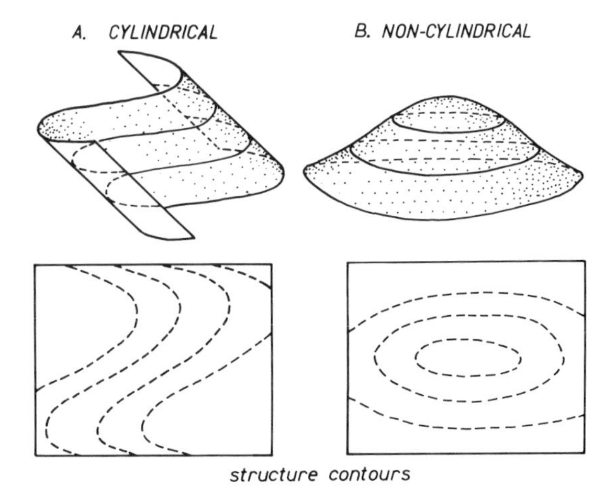

FIG. 3.20 Structure contour patterns of cylindrical and non-cylindrical folds.

(Fig. 3.21A). Lines can be drawn on any contour map marking the 'valley bottoms' and 'the brows of ridges' in the folded structure. These lines are called the *trough lines* and *crest lines* respectively. The recognition of these antiformal crests and synformal troughs forms a useful preliminary step in the interpretation of a structure contour map. (Fig. 3.21B).

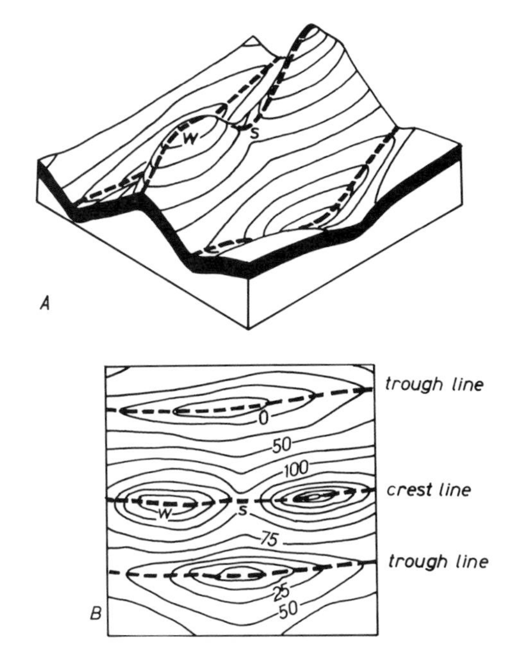

FIG. 3.21 A folded surface.

Exercise

Bend a piece of card into an angular fold (Fig. 3.22A). Tilt the card so that the fold plunges. With chalk, sketch in the run of structure contours on the folded card and draw, on a map, how these contours will appear as seen from above (Fig. 3.20A). Repeat this for other angles of plunge to investigate how the plunge affects both the shape of the contours and their spacing. Tilt the fold enough to make one limb overturned, i.e. to make the fold into an overturned fold. Draw structure contours for the fold. Note that the contour lines for the one limb cross with those of the other limb (Fig. 3.22C). This type of pattern signifies that overturned folds are 'double-valued' surfaces; that is the surface is present at two different heights at the same point on the map.

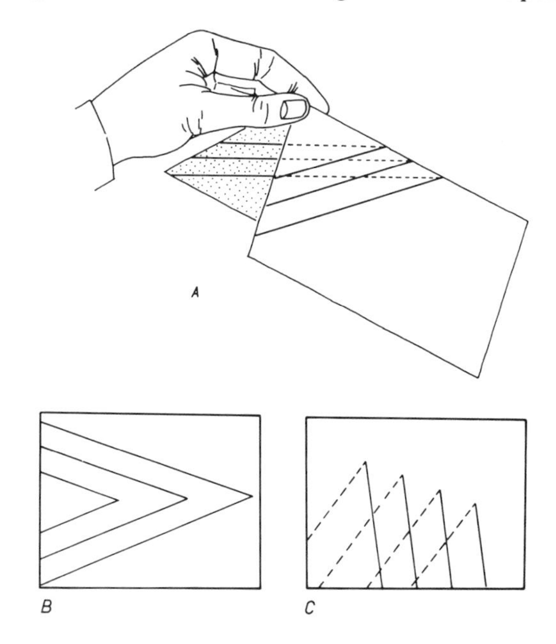

FIG. 3.22 Visualizing the structure contour patterns of plunging folds.

3.10 Determining the plunge of a fold from structure contours

Figure 3.23 shows a plunging fold and a map of its structure contours. If the structure contours are given, the crest (or trough) line can be drawn in. For a cylindrical fold this line is parallel to the hinge line, so that its plunge is measured straight from the map. It is the trend of the crest/trough line in the 'downhill' direction. The direction of plunge is shown on maps by an arrow (Fig. 3.23B; also see 'Geological map symbols'). The angle of plunge is calculated by solving the triangle in Fig. 3.23A. Note that the contour spacing mentioned is the separation of the contours measured in the direction of plunge.

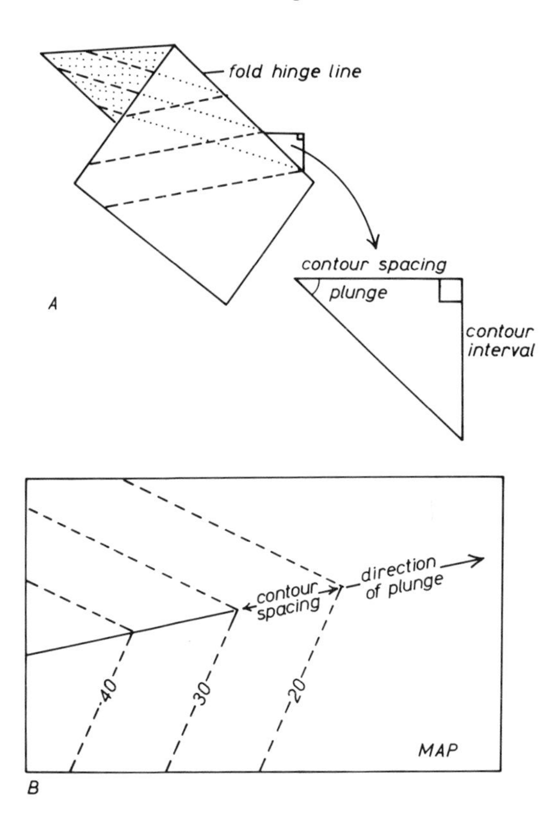

FIG. 3.23 Calculation of the fold plunge from structure contours.

3.11 Lines of intersection of two surfaces

For angular folds with planar limbs the trough/crest line can be considered to be the line of intersection of one limb with the other. Therefore the above method to calculate the fold plunge can be applied to calculate the plunge of the line of intersection of any two surfaces (Fig. 3.24). If both surfaces are planar (dip uniformly) their line of intersection will be straight (plunge uniformly).

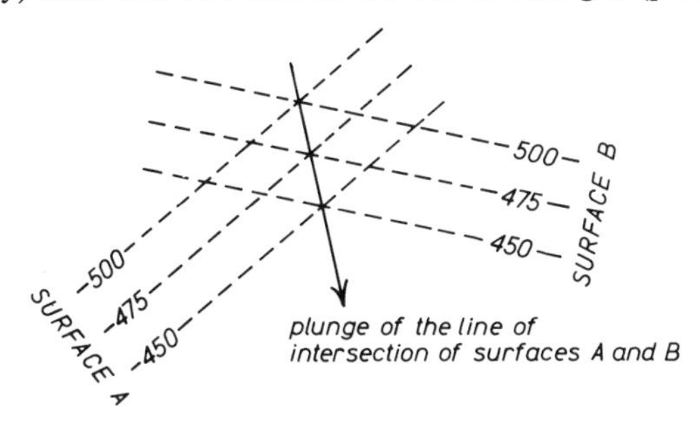

FIG. 3.24 Calculation of the line of intersection of two planar surfaces.

3.12 Determining the plunge of a fold from the dips of fold limbs

The previous calculation of fold plunge from structure contours also allows the position of the fold crest or trough line to be determined. If, instead of structure contours, two limb dips only are known, the plunge can still be calculated.

Worked example

One limb of a fold has a dip of 207/60 and the other limb dips 100/30. Determine the plunge of the fold axis.

As the limbs are not parallel they will intersect somewhere to give a line of intersection which is parallel to the fold axis. Let X be a point on the line of intersection of height h metres. The structure contours of elevation h metres for each of the limbs must intersect at X. These can be drawn parallel to the strike of each limb (Fig. 3.25A). Adopting a convenient scale, use the known angles of dip and the equation in Section 2.9 to calculate the position of structure contours for an elevation of $(h-10)$ metres. These intersect at a point Y, which like X is a point on the line of intersection. The height difference between X and Y is 10 m and their horizontal separation is 20 m. The triangle in Fig. 3.25B can be solved for the angle of plunge. The fold axis thus plunges at 26°. The direction of plunge is 138°.

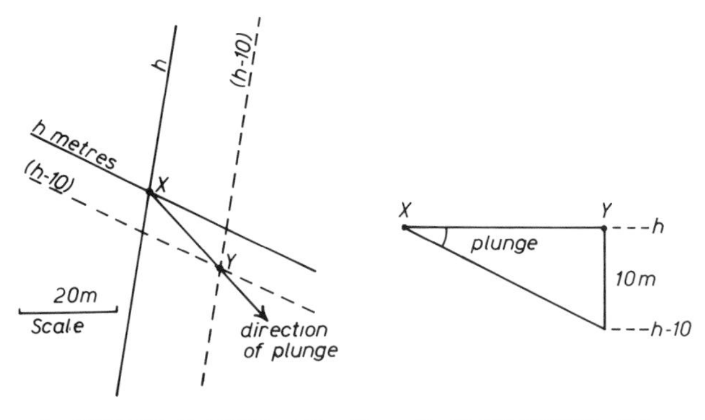

FIG. 3.25 Calculating the plunge of a fold axis.

It can be readily shown by means of a figure similar to Fig. 3.24 that if two portions of a folded surface have the same angle of dip then the direction of plunge bisects the angle between their directions of dip. Also, if the dip on any part of a folded surface is vertical then the strike of that part is parallel to the direction of plunge. These rules sometimes provide a direct method of deducing the direction of plunge of a fold from the dip symbols shown on a geological map. The angle of plunge is obtained by looking for a dip direction which is, as near as possible, parallel to the fold plunge direction. The angle of dip there will be approximately equal to the angle of plunge. No part of a folded surface can dip at an angle less than the angle of plunge of a fold. It is important to remember that these methods apply only to cylindrical folds.

3.13 Sections through folded surfaces

We usually observe folds not as completely exposed undulating surfaces, but in section, as they appear exposed on the surface of a field outcrop or on the topographic surface. In other words our usual view of folds tends to be two-dimensional, similar to folds viewed in cross-sections. Figure 3.30 shows the outcrop pattern yielded by plunging folds.

Exercise

Bend a piece of card into a number of folds. With scissors, cut planes through the folded structure. On each cut section (Fig. 3.26) note carefully (a) the tightness, (b) the asymmetry, and (c) the curvature of the folds seen in oblique section. This exercise demonstrates the importance of the orientation of the slice through the fold for governing the fold shape one observes.

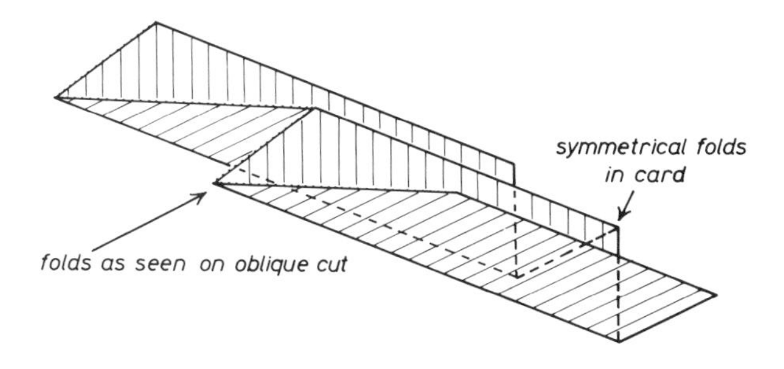

FIG. 3.26 Classroom experiment to illustrate the importance of the section effect on a fold's appearance.

3.14 The profile of a fold

The fold profile is the shape of a fold seen on a section plane which is at right angles to the hinge line. The fold profile or *true fold profile* is important since only this two-dimensional view of the fold gives a true impression of its tightness, curvature, asymmetry, etc. It corresponds to our view of the fold as we look down the fold hinge line.

3.15 Horizontal sections through folds

Folds displayed on a map often present us with a more or less horizontal section through a fold. In order to be able to interpret the observed geometry in terms of the three-dimensional shape of the fold, the technique of *down-plunge viewing* is useful. This entails oblique viewing of the section so that the observer's line of sight parallels the plunge of the structures (Fig. 3.27). The view so obtained corresponds to a true profile of the fold structure.

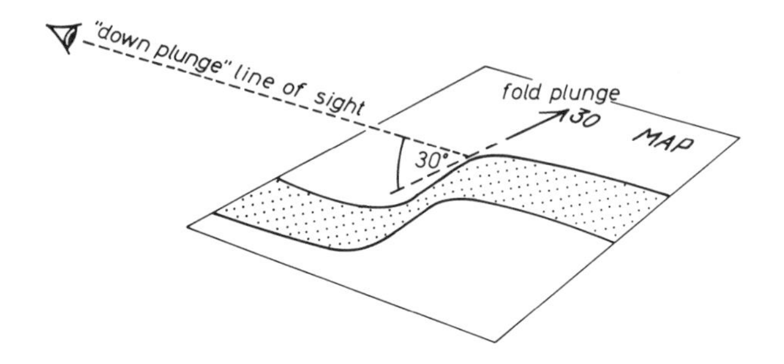

FIG. 3.27 The 'down-plunge' way of viewing a geological map allows the true shape of a fold structure to be seen.

Worked example

The map in Fig. 3.28A shows a fold exposed in an area of relatively flat topography. Use down-plunge viewing to answer the following:

(1) Is the fold an antiform or a synform?
(2) How does the fold classify in the tightness classification (Section 3.4)?
(3) Is the fold approximately similar or approximately parallel (Section 3.8)?

View the map looking downwards towards the SW so that your line of sight plunges at 27° towards 250°. The view so obtained is shown in Fig. 3.28B. It reveals the fold to be an antiform which classifies as 'close' (interlimb angle = 31°) and its shape is approximately parallel (constant stratigraphic thickness).

Providing the direction of plunge is known, the apparent direction of closure on a horizontal surface indicates whether a fold is antiformal, synformal or neutral (Fig. 3.29). The last example serves to emphasize how oblique sectioning produces an apparent shape which can differ significantly from a true profile. On maps of more or less horizontal surfaces this effect is most marked when plunges are low.

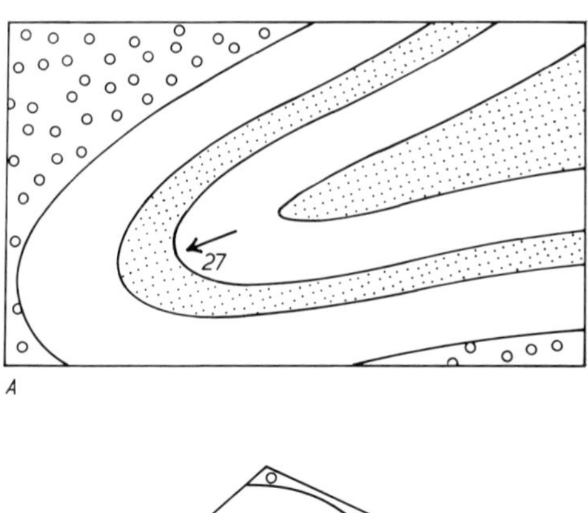

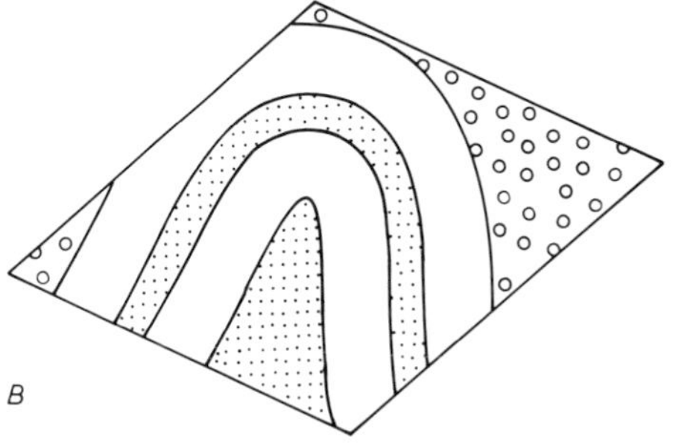

FIG. 3.28 An example of the 'down-plunge' method of viewing a map.

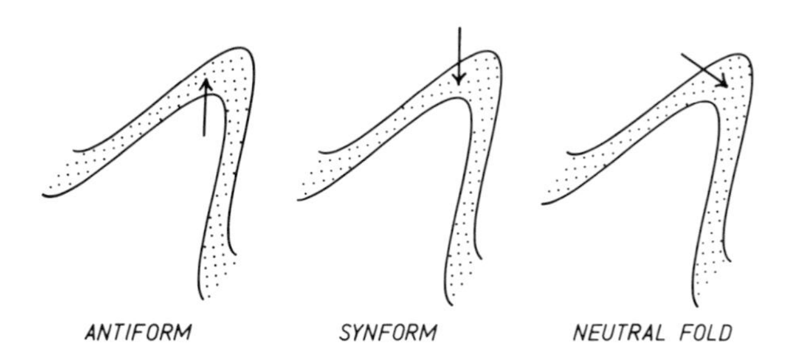

FIG. 3.29 The outcrop pattern produced by three folds with different directions of closure. The arrows show the plunge directions.

FIG. 3.30 **A**: An upright synform (also a syncline) exposed on a wave-cut platform at Welcombe, Devon (Upper Carboniferous sandstones and shales). The synform plunges towards the sea.

FIG. 3.30 **B**: Plunging upright folds, Svartisen, Norway. The hand points down the hinge line of a synformally folded quartz-rich layer in the gneisses. Note the way the fold appears to close on the horizontal and steep parts of the outcrop surface.

3.16 Construction of true fold profiles

This is a formalization of the down-plunge viewing method. Oblique sectioning of folds on non-profile planes produces a distortion of the shape which can be most easily visualized by reference to a plunging circular cylinder, say a bar of rock (of the candy variety!) The cylinder of diameter D plunges at an angle Θ and will appear as an ellipse on a horizontal section plane (Fig. 3.31A). This ellipse is a distorted view of something which in true profile has the shape of a circle. The shape of the ellipse shows that distortion consists of a stretching in the direction of plunge such that a length D appears to have length $D/\sin\Theta$ (Figs. 3.31B and 3.31C).

The construction of the true profile involves taking off this distortion. Place a rectangular grid on the map (Fig. 3.31B) with one axis (x) parallel to the direction of plunge and where the scales in the x and y directions are in the ratio $1:\sin\Theta$. The coordinates of points on the fold outline referred to this grid are replotted on a square grid (Fig. 3.31D). This shortening of dimensions parallel to the plunge compensates for the stretching brought about by oblique sectioning.

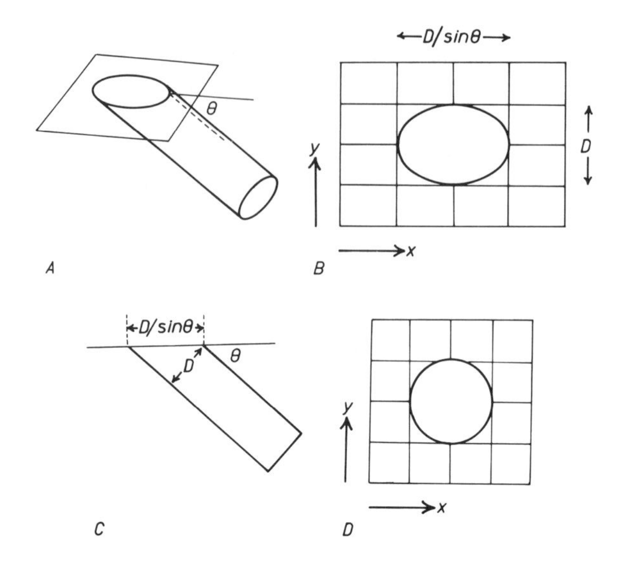

FIG. 3.31 Construction of a true fold profile.

3.17 Recognition of folds on maps

The interpretation of the structure of an area represented on maps is greatly assisted if the patterns produced by fold structures can be directly recognized. Students should use every opportunity to practise their skills in this aspect of geological map reading.

In Chapter 2 the sinuosities produced by the effect of topography on uniformly dipping beds were discussed. Many of the shapes produced look superficially like sections through folds (see Fig. 3.32A).

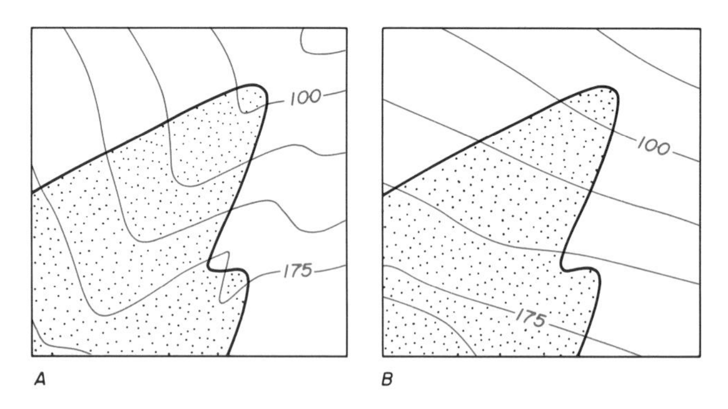

FIG. 3.32 **A**: Outcrop pattern related to topography.
B: Outcrop pattern related to folding.

These outcrop patterns are characterized by V-shapes in parts of the map where V-shapes are shown by the topographic contours. On the other hand outcrop patterns due to folding possess definite curvatures which cannot be related to a corresponding pattern in the topographic contours (Fig. 3.32B). Fold-like sinuosities of geological contacts in a part of a map where the terrain consists of a uniform slope must be the expression of folding.

Worked example

The map in Fig. 3.33 shows the outcrop of a thin shale. The highly curved outcrop pattern shows four strongly curved parts (*A*, *B*, *C* and *D*). Explain the reason for the fold-like shapes.

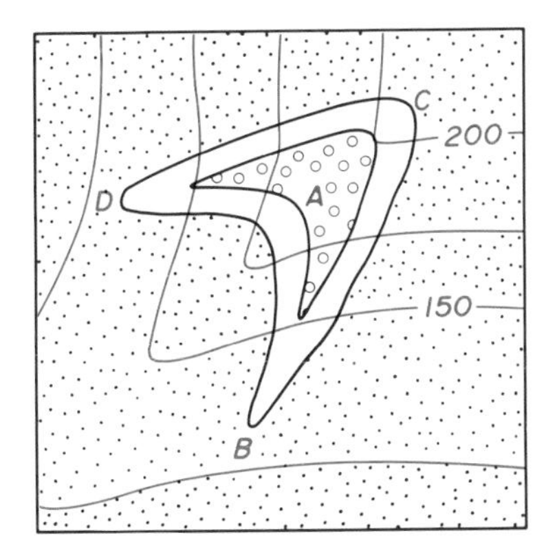

FIG. 3.33 Structural interpretation of outcrop pattern.

The bends at *A* and *C* occur on a ridge defined by a V-shaped arrangement of topographic contours. These curves are therefore due to the terrain effect and are not an expression of folding. The curvatures at *B* and *D* are not related to a particular topographic feature, and occur on fairly uniformly sloping valley sides. *B* and *D* therefore represent folds.

3.18 Hinge points and axial surface traces

The points labelled *B* and *D* in Fig. 3.33 are points of maximum curvature in a fold and can be referred to as *hinge points*. Strictly speaking, the curvatures we observe on the map are prone to the oblique sectioning effect dealt with in Section 3.13. This means that the point of maximum curvature of the map may not lie exactly on the real hinge line of the fold. In practice, though, this problem is not important, especially for fairly angular fold profile shapes.

The *axial surface trace* or axial trace is the line of intersection of the axial surface and the ground surface. In three dimensions the axial surface cuts through the geological contacts along lines which are fold hinge lines (see Fig. 3.8). On a geological map the axial surface trace runs as a line across geological contacts crossing them at the hinge points.

Worked example

Draw the axial surface trace for the folded schist/quartzite contact on the map (Fig. 3.34A).

Using the reasoning of Section 3.17 points *a* and *b* can be identified as hinge points (*c* and *d* are topographic V's). Starting at point *a* , the axial trace must here cut the contact, passing from the schist into the quartzite. The trace of the axial surface continues within the quartzite and may only 'escape' from the quartzite outcrop if another hinge point is present allowing it to cut through a geological boundary. Such a hinge point exists at *b*, and it is here that the axial surface trace passes from the quartzite into the schist (Fig. 3.34B).

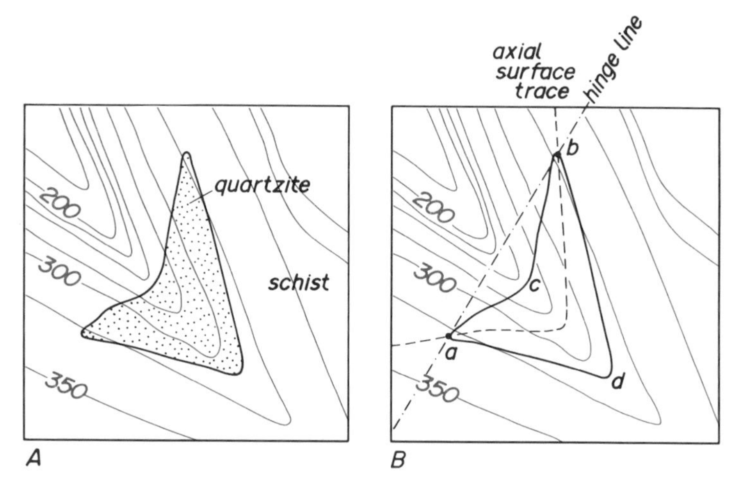

FIG. 3.34 Recognizing hinge points, axial surface traces and hinge lines.

The axial surface trace marks the outcrop position of a surface, and even though this surface is imaginary the form of the trace and how it reacts to topography can be interpreted like any other surface. Structure contours for the axial surface can be drawn (Section 2.15) and the V-rule (Section 2.14) can be applied to determine its dip. For example, axial surface traces which follow closely the trend of the topographic contours indicate recumbent folding and straight axial traces characterize upright folds.

Axial traces serve to divide a map into strips corresponding to individual fold limbs.

3.19 Constructing hinge lines on maps

In a sequence of layered rocks affected by a single set of folds each axial surface never crosses a geological contact surface more than once. This fact means that all hinge points in an axial surface involving a particular geological boundary must be points belonging to the same hinge line. Hinge lines are drawn on geological maps by joining two or more points in an axial surface trace where a particular contact is intersected. In Fig. 3.34B hinge points *a* and *b* have been joined to form a hinge line for the quartzite/schist contact. This is permissible as both points lie on the same axial surface trace and on the same contact. The hinge line has been drawn straight, but had more hinge points been available it would have been possible to check whether the hinge line is in fact straight. In the absence of additional hinge points a constantly plunging hinge line is assumed between *a* and *b*. The angle of plunge is calculated by the height difference and horizontal distance separating the points *a* and *b* (see Section 3.10). As *b* is the lower of the two hinge points, the plunge direction is the compass direction a—>b, which is 031°.

3.20 Determining the nature of folds on maps

An essential part of geological map reading is the ability to deduce whether a given fold closure displayed on a map is antiformal or synformal. Use the procedures described above to identify fold closures and assess the plunge of their hinges. Select a part of the map where the ground slope is opposed to the plunge of the folds. Visualize the fold shape that could be seen on the hillside from a low vantage point on the ground and facing up-slope. In the case of the map in Fig. 3.35A the reader should imagine the view from a spot in the stream looking northwards and up at the hillside which slopes towards the observer. The quartzite/schist contact crosses the stream to the viewer's left and runs uphill until a height of about 325 m is reached, and then suddenly starts to descend, ending up in the stream to the observer's right (Fig. 3.35B). The structure clearly closes upwards and is therefore an antiform.

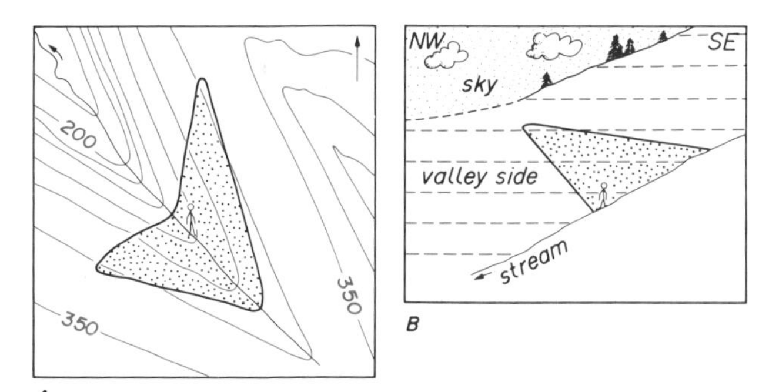

FIG. 3.35 Deducing if a fold on a map is an antiform or synform.

3.21 Cross-sections through folded areas

The procedure for preparing a cross-section is best explained by an example.

Worked example

Draw a vertical section across the map in Fig. 3.36A between points *X* and *Y*. This is tackled in several stages:

(1) distinguish true fold hinge points from topographic effects (Section 3.17);
(2) join these hinge points to form axial surface traces (Section 3.18);
(3) join hinge points to form hinge lines (Section 3.19).

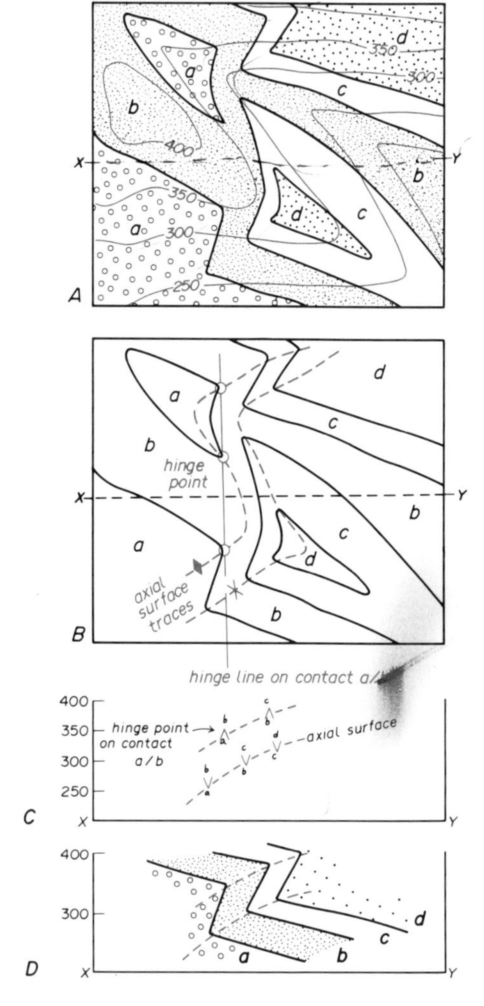

FIG. 3.36 Drawing a cross-section through a folded area.

After drawing in the topography the shapes of the axial surface traces are constructed in the cross-section (Fig. 3.36C). This is most easily done by transferring the hinge points to the section by interpolating their heights from where hinge lines intersect the plane of section (Fig. 3.36B). The axial surface traces are then drawn in the section by joining the hinge points. The final stage (Fig. 3.36D) involves drawing the folded geological boundaries. These form the limbs of the folds and can be considered as the portions of the fold linking hinge points from adjacent axial surfaces. If the more exact shape of the limbs is required additional data can be obtained from structure contours constructed for the geological boundaries on the limbs.

3.22 Inliers and Outliers

Some geological boundaries on a map define a closed loop, so that rocks belonging to one unit completely enclose those of another. The outcrop pattern involving older rocks being surrounded by younger ones is given by the name *inlier*. The reverse relationship, with younger rocks completely encircled by older rocks, is present in an *outlier*.

These outcrop patterns can be produced either by erosion (e.g. the outlier 2 km WNW of Winchcombe on the map in Problem 2.7) or by folding (especially periclinal folding). The pattern in Fig. 3.35 is due to a combination of folding and erosion. If we knew that the quartzite was older than the schist, the structure in Fig. 3.35 would be an inlier.

Answer to question set in section 3.3:

Vertical folds, reclined folds and, possibly, recumbent folds are examples of neutral folds.

PROBLEM 3.1

The numbers on the map give the depths (below sea level) to the top of the Wittekalk Limestone.

1. Draw structure contours for the top of the limestone.
2. Explain why the structures cannot be interpreted as cylindrical folds.
3. Label the crest lines and trough lines on the map.
4. Give names to the various periclinal folds present.
5. Calculate the dip direction and dip at points *p* and *q* on the top of the limestone.

PROBLEM 3.2

The photograph, taken from the air, in the Zagros Mountains, Iran, shows folded rocks.

Draw a sketch map of the area to show the geological structure. Use on your map as many as possible of the symbols given on the list at the beginning of this book (see 'Geological map symbols').

PROBLEM 3.3

For each map:

(a) draw the fold axial surface trace (axial trace),
(b) determine the plunge of the fold hinge and show this on the map with the appropriate symbol (see list of symbols at beginning of book),
(c) determine the dip of the axial surface and show this on the map with the appropriate symbol,
(d) classify the fold according to the scheme based on fold orientation (Fig. 3.9).

PROBLEM 3.4

Each map shows folded beds for which the younging direction is known. Classify the folds:

(a) as antiforms, synforms or neutral folds;
(b) as anticlines or synclines;
(c) as upward-facing, downward-facing or sideways-facing folds.

The answers to (a), (b) and (c) are not independent. By this we mean, for instance, that a knowledge of the facing direction (upward/downward) and the direction of closure (antiform/synform) allows us to decide whether a fold is an anticline or syncline. Devise a rule explaining the interdependence of the three classifications. (A series of sketches showing the various possible combinations will help to clarify your thinking.)

PROBLEM 3.5

The map shows the geology of an area in the province of Namur in Belgium.

The distribution of the rocks and the shape of their outcrops are structurally, rather than topographically, controlled.

The fold structures in the area, which are of Variscan age, are all upward-facing.

Working on a tracing paper overlay to the map:

(a) mark the axial surface traces of anticlines and synclines,
(b) deduce the approximate direction of plunge from the direction of fold closure and show this with plunge arrows on the various parts of the map (see list of symbols), and
(c) show the location of an example of an inlier and an example of an outlier.

Analyze the regional variation of fold plunge by locating the position of plunge culminations and plunge depressions.

Comment on the style of the folds responsible for the elongate oval outcrop patterns.

PROBLEM 3.6

By consulting Fig. 3.17 draw hypothetical maps showing the outcrop patterns of periclinally folded beds, exposed in an area of low relief. On these sketch-maps use symbols to show the dip of beds, as well as course of the geological boundaries. Does each type of periclinal fold produce a characteristic outcrop pattern?

The accompanying map is a geological map of part of Derbyshire.

Identify and mark the hinge points of folds. Establish whether these folds are antiforms or synforms. Draw the axial traces of antiformal and synformal folds.

Deduce the approximate direction of plunge of the various folds present.

Where do the following types of periclinal folds occur on the map: canoe-, whaleback-, shoe-horn-, and saddle-shaped folds?

Why do the sandstone beds (shown with stippled ornament) vary so much in their width of outcrop?

PROBLEM 3.7

The accompanying map shows Palaeozoic rocks exposed in a part of Gloucestershire.

Describe the most important structure in the area and, after calculating the fold plunge, construct a *true profile* of the structure.

From the true profile, measure the interlimb angle of the fold. What name is given to a fold with this interlimb angle (see Section 3.4)?

Examine the thicknesses of the rock units around the fold. Is the fold closest to a parallel or similar type? Measure the thickness of the Carboniferous Limestone. Is it correct to say that the Carboniferous Limestone is thicker in the region of the fold hinge?

PROBLEM 3.8

Practise using the terms explained in this chapter to describe, as fully as possible, the geometry of the folds in the photographs. For example, refer to the tightness, orientation, layer thickness variation, curvature, etc. of the folds. All photographs are of vertical outcrop surfaces.

Draw sketches of the folds showing, where appropriate, the axial traces, hinge points, inflection points, trace of the enveloping surface, fold limbs, etc.

Photographs A and C show structures in metasedimentary gneisses, B shows folded quartz veins (Nordland, Norway), D shows folded turbidites, Crackington, north Cornwall.

PROBLEMS 3.9 to 3.11

(in order of increasing complexity)

On each map:

(a) mark hinge points, antiformal and synformal axial traces and hinge lines;
(b) determine the plunge and plunge direction of the fold hinge lines;
(c) deduce the general attitude (dip) of the fold axial surfaces;
(d) construct a vertical cross section along the line *X-Y*,
(e) give a full description of the characters of the folds (e.g. tightness, symmetry, bed thickness variations, etc.).

PROBLEM 3.1

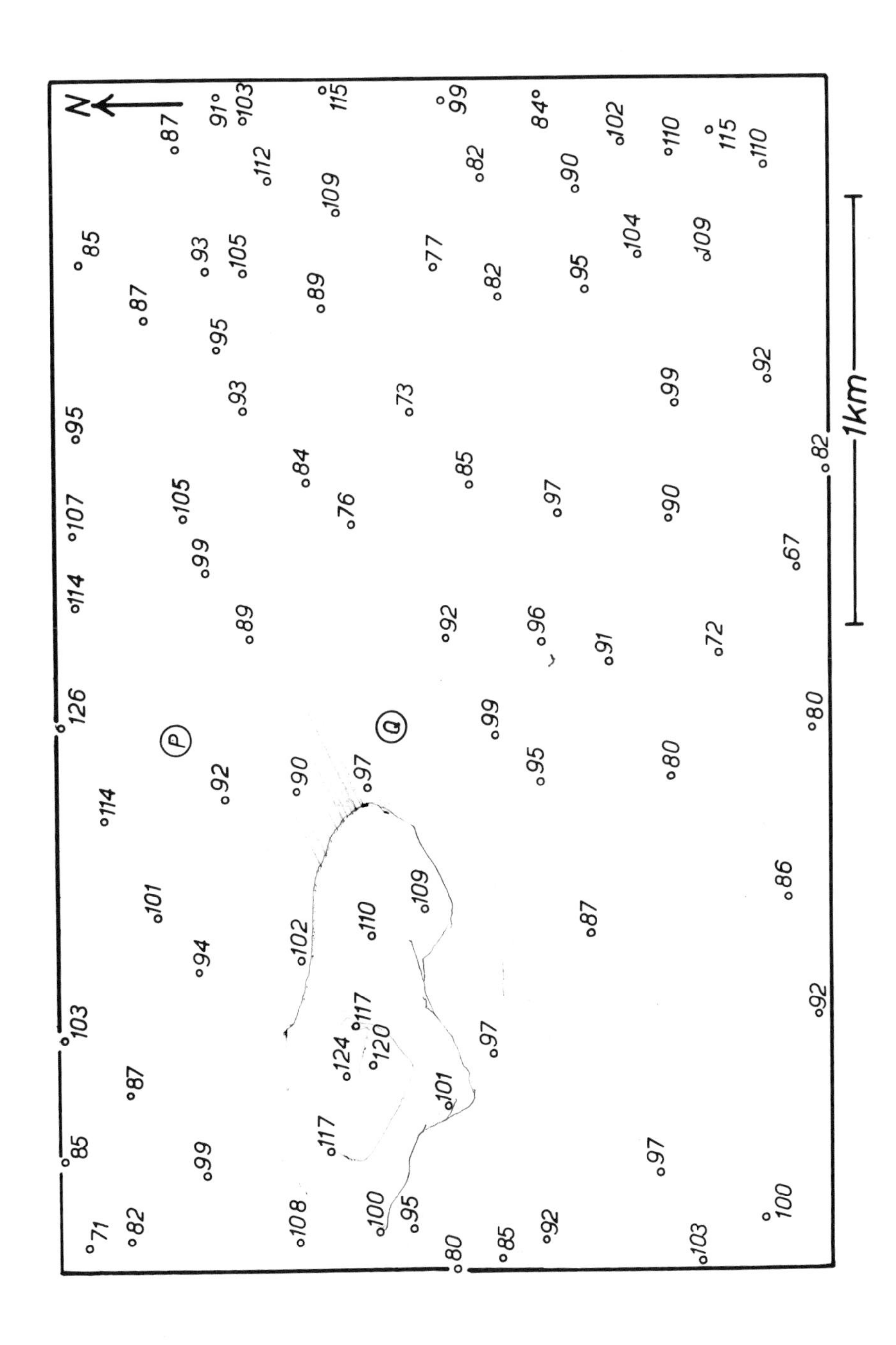

PROBLEM 3.2

(photo: courtesy Aerofilms)

PROBLEM 3.3

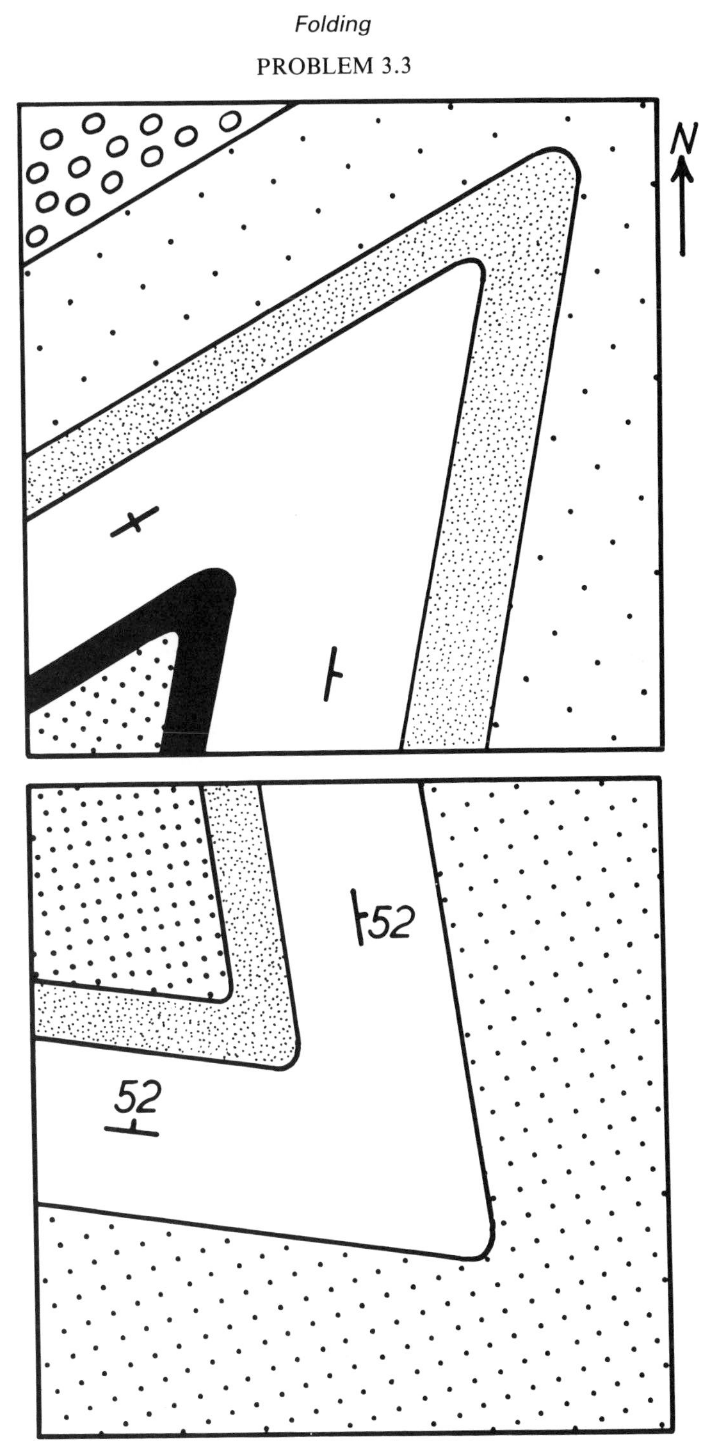

PROBLEM 3.4

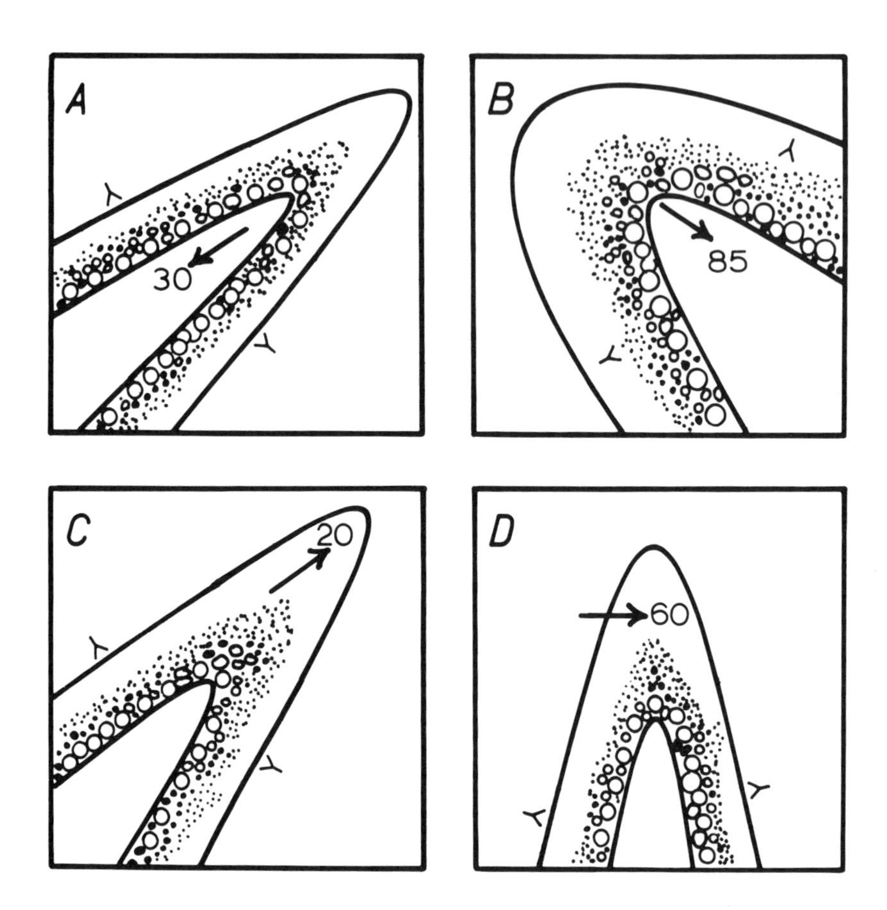

PROBLEM 3.5

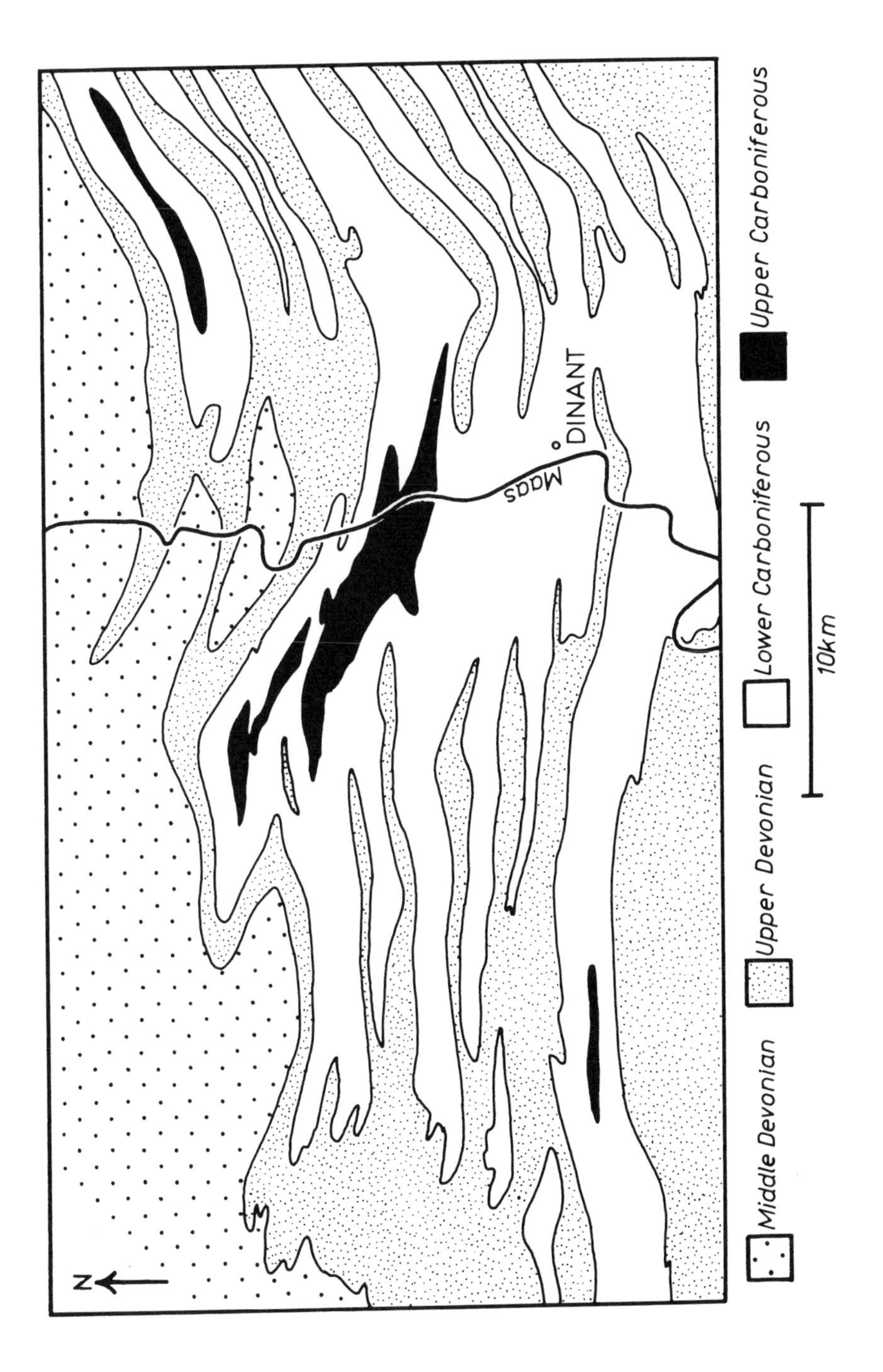
N
DINANT
Maas
Middle Devonian
Upper Devonian
Lower Carboniferous
Upper Carboniferous
10km

PROBLEM 3.6

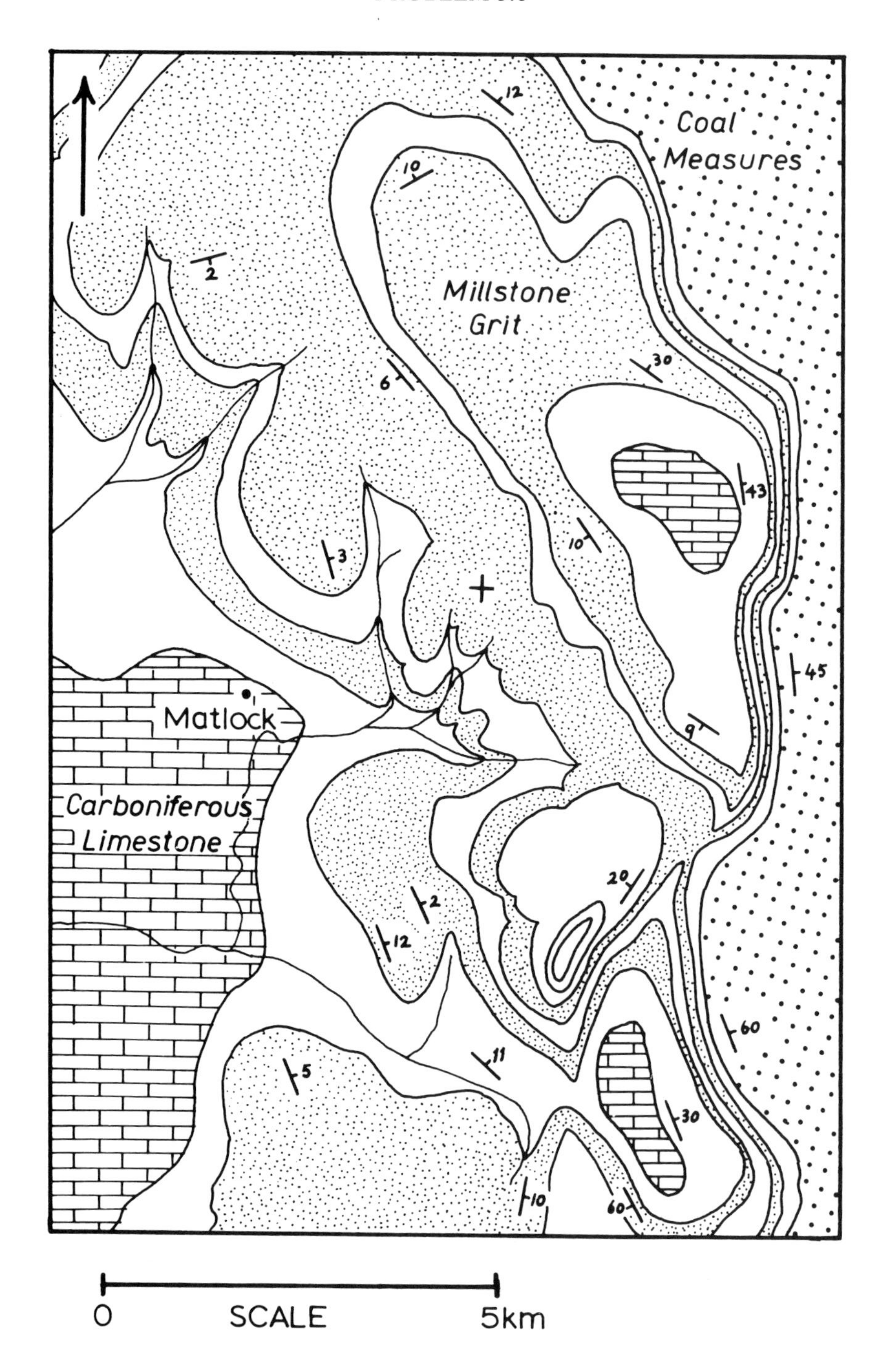
Coal Measures
Millstone Grit
Matlock
Carboniferous Limestone
12
10
2
6
30
43
10
3
45
9
20
2
12
60
11
5
30
10
60
0
SCALE
5km

PROBLEM 3.7

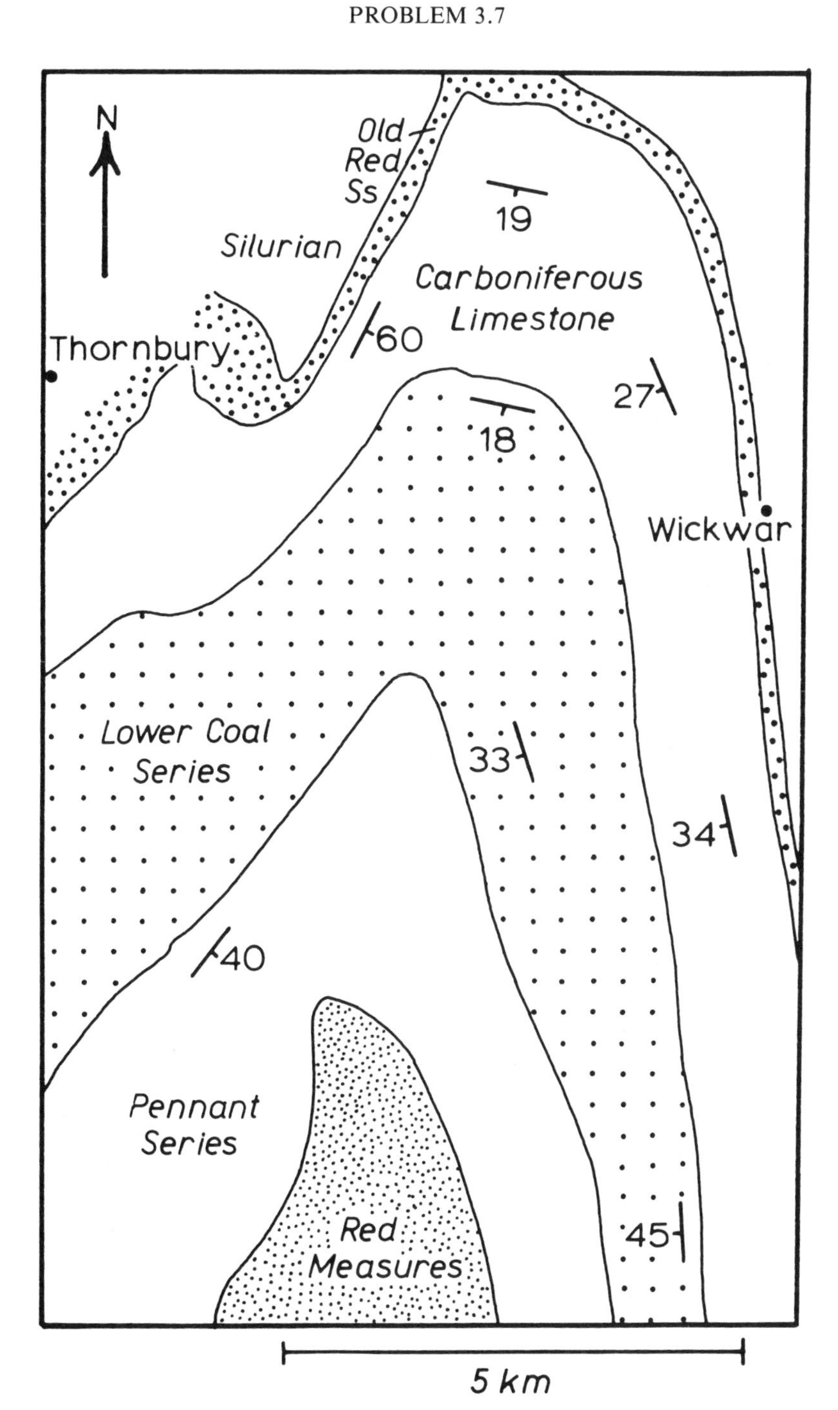
N
Old
Red
Ss
Silurian
19
Carboniferous
Limestone
60
Thornbury
27
18
Wickwar
Lower Coal
Series
33
34
40
Pennant
Series
Red
Measures
45
5 km

PROBLEM 3.8 (A & B)

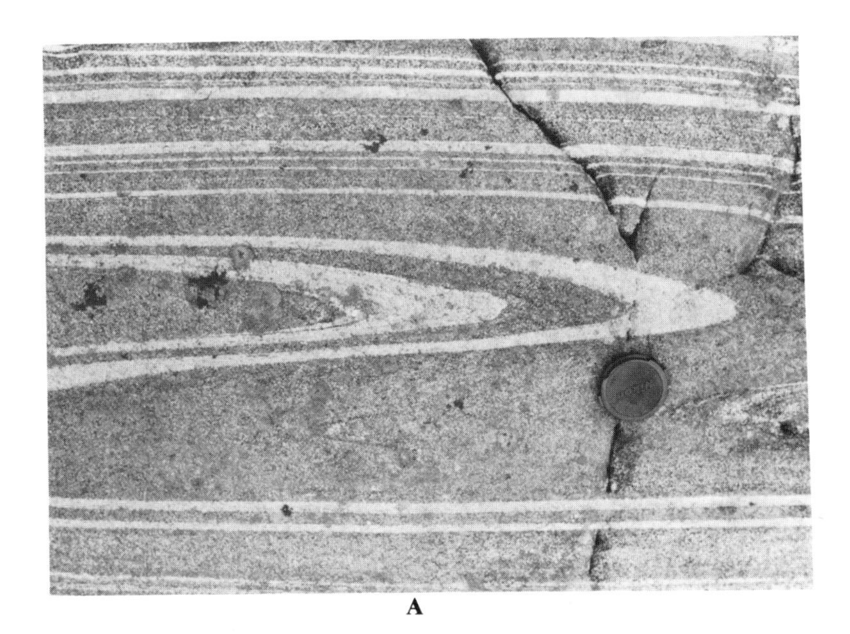

A

B

PROBLEM 3.8 (C & D)

C

D

PROBLEM 3.9

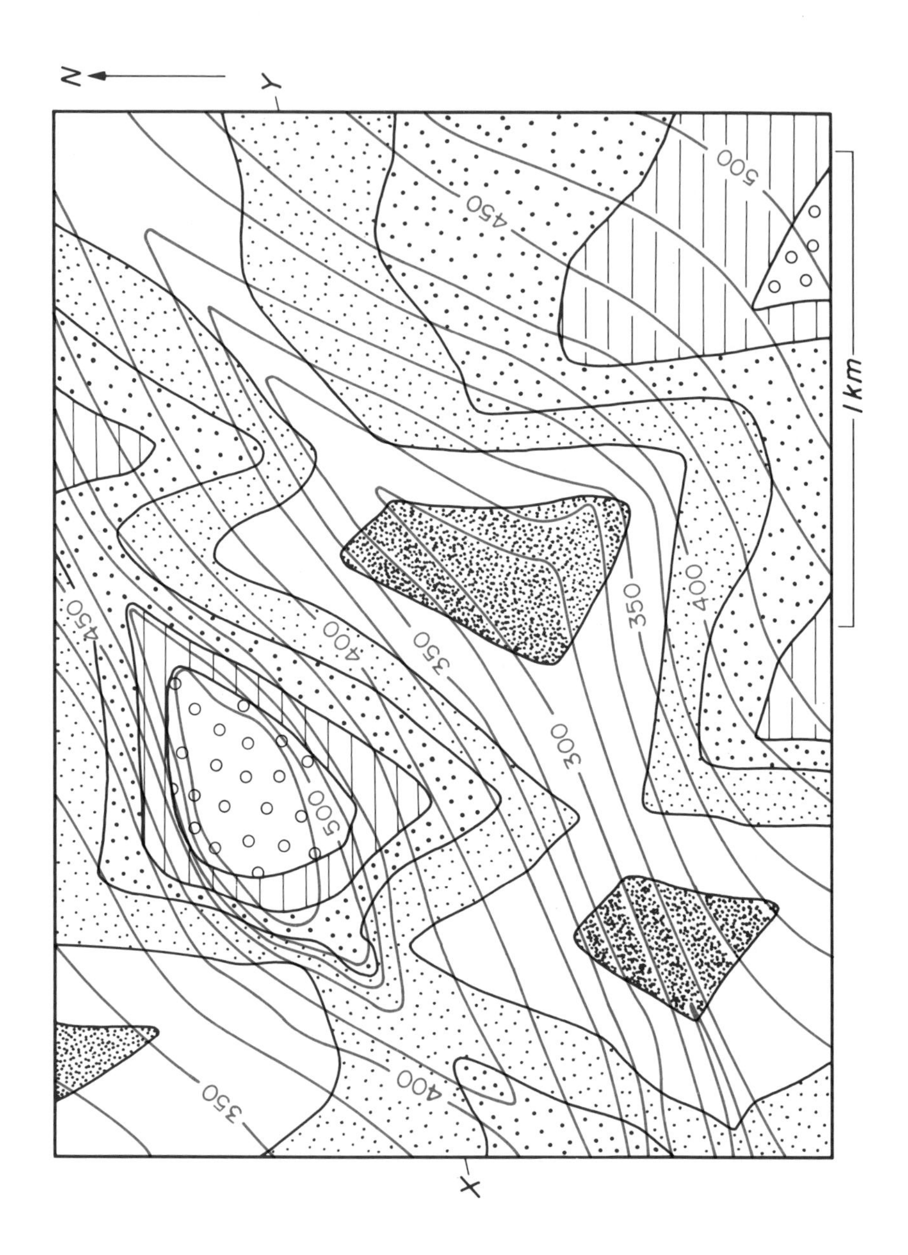

PROBLEM 3.10

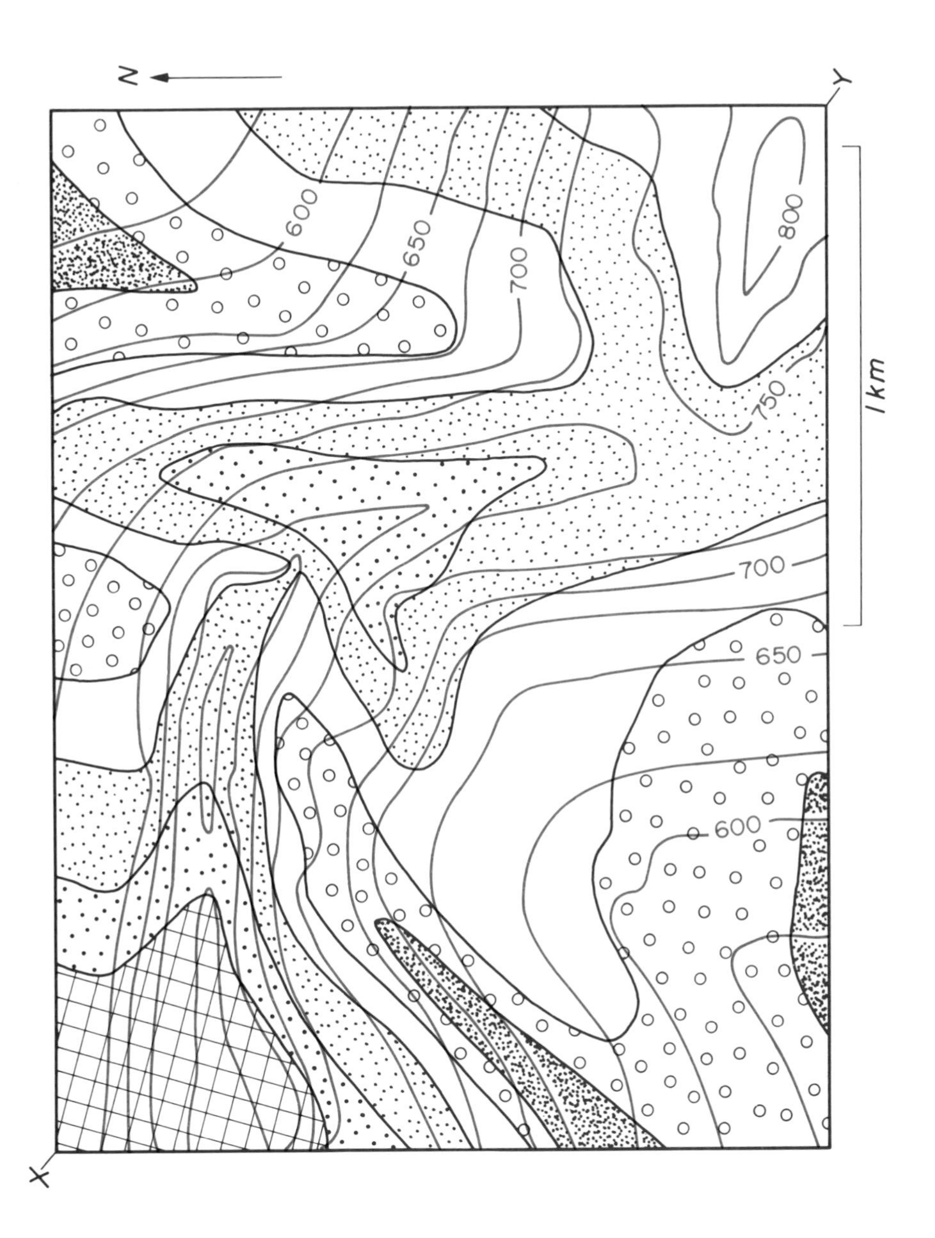

PROBLEM 3.11

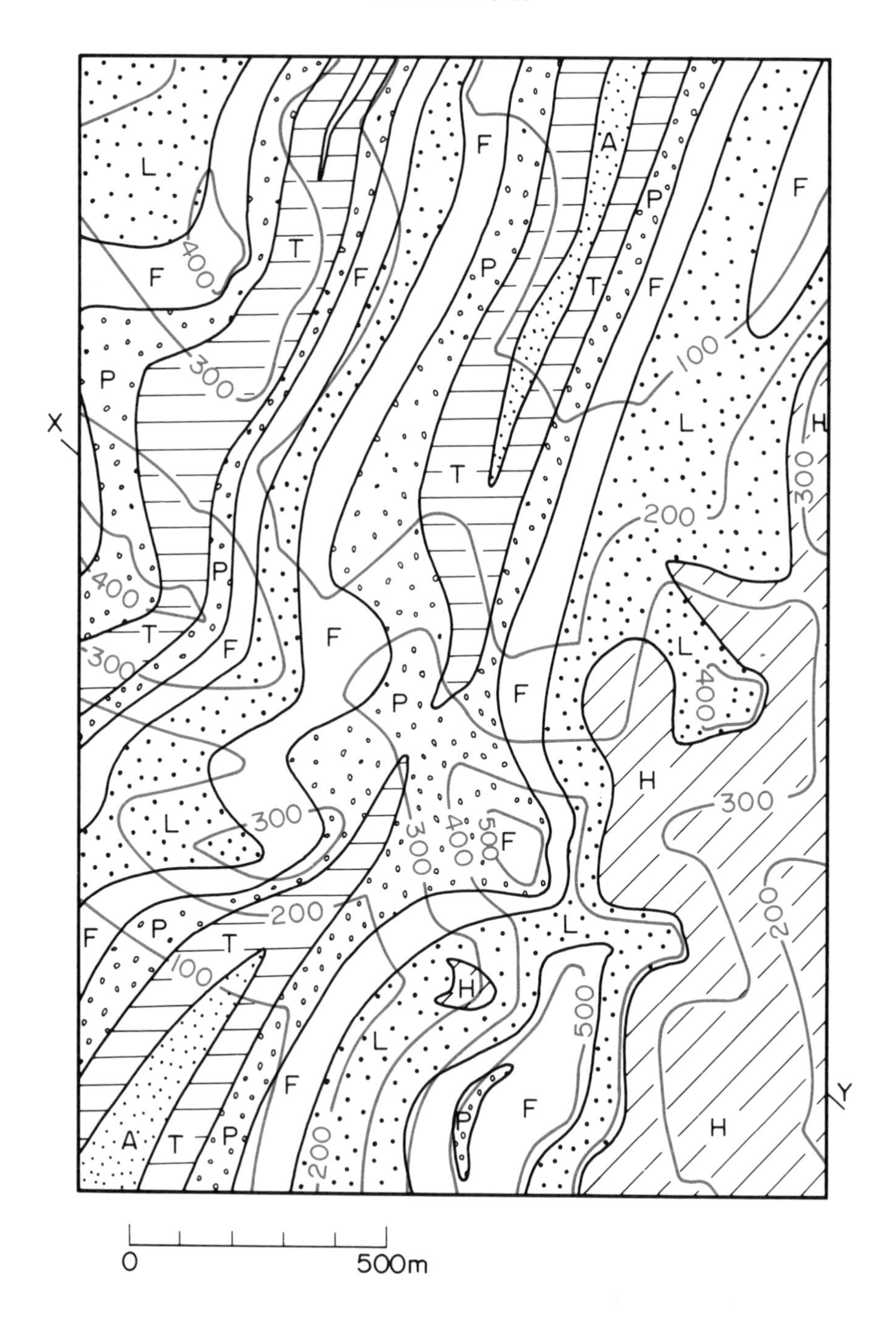

4
Faulting

FOLDS and faults are both structures produced by the deformation of rocks. Folds are structures where layering is deformed without breaking so that the layering surfaces are curved but continuous (Fig. 4.2A). Faults represent a different type of response by the rocks to the stresses imposed on them. Faults are fracture surfaces along which appreciable displacement of the layering has taken place (Figs 4.1 and 4.2B). Faults, unlike folds, can be thought of as structural discontinuities.

FIG. 4.1 A fault is a rock fracture across which displacement has occurred. Evidence of the motion on the fault comes from displaced marker layers.

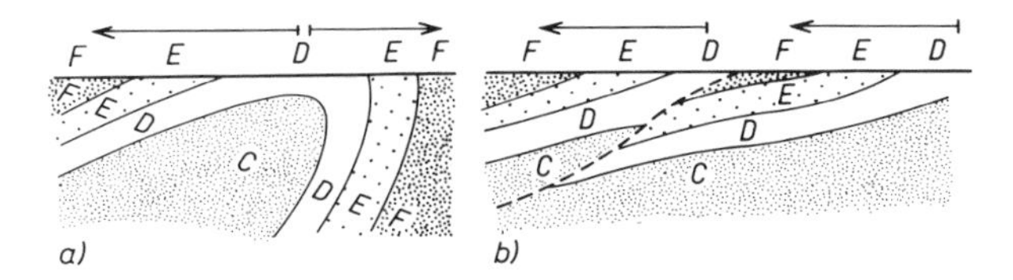

FIG. 4.2 Sequences of rocks observed on line traverses through folds (**A**) and faults (**B**).

Figure 4.2 illustrates an important difference between faults and folds. In each sketch a stratigraphy consisting of beds *A* to *F* is being deformed. In each cross-section a horizontal line traverse reveals a characteristic sequence of encountered layers. In the case of folding the sequence is

F E D E F
← →
sequence sequence reversed

With the fault the sequence is

F E D F E D
← ←
sequence sequence repeated

Folds can give rise to a duplication of a set of beds where the order in which they appear shows reversals. It should be noted that the axial surface marks the place where the sequence reverses. In the example of faulting, we note that the sequence is not reversed but repetition of the same sequence takes place. The fault plane marks the place where the sequence starts to repeat itself. The fault in Fig. 4.2B places units *D* and *F* next to one another. The contact *D*/*F* is not a normal contact because *D* and *F* do not normally occur as adjacent units. The fault forms what is called a *tectonic contact* between units *D* and *F*. Not all faults give rise to repetition of encountered beds, but some are instead discontinuities at which certain beds are omitted. These differences between folds and faults allow the structures to be distinguished in line transects (e.g. boreholes) and are summarized below.

Table 4.1

	Folds	Faults
Linear traverses through structure show :	Continuous sequences, with reversals possible	Discontinuous sequences with possible duplication and omissions of units; no reversals of sequences

4.1 Fault planes

With a fault, the surface along which displacement has taken place is the *fault plane*. Structure contours can be used to describe the form of fault planes and these are interpreted in the same way as structure contours drawn for any surface (see Chapters 2 and 3). The inclination of a fault plane is described by its dip. Faults with dips greater than 45° are sometimes referred to as *high-angle faults*; those with dips less than 45° are *low-angle faults*. The rocks which lie above a fault plane make up the *hanging wall*, those below the *foot wall* (Fig. 4.3).

Faults displace planar surfaces such as bedding planes. Any geological surface (such as a bedding plane) is truncated by a fault along a line called the *cut-off line*. On a map or cross-section the point where the surface is crossed by the fault line is a *cut-off point*. Lines can be drawn on maps showing the course of the cut-off lines in the subsurface. These lines, being the lines of intersection of two surfaces, are calculated from the structure contours of each surface (see Sections 3.10 and 3.11).

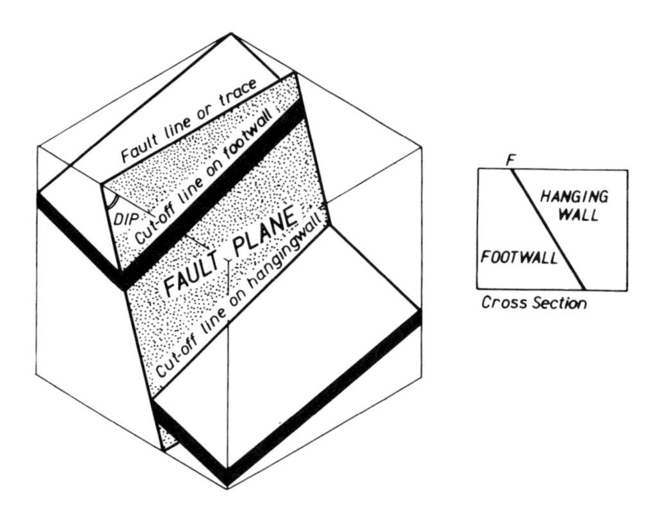

FIG. 4.3 Fault terminology.

Worked example

Determine the cut-off line for the top of the sandstone bed in Fig. 4.4A.

From the given dips, structure contours for heights relative to that of the ground surface are drawn for both the fault plane and the top of the sandstone bed. The intersection of the respective contours gives the line of intersection (the cut-off line). This line passes through the visible cut-off point on the map (Fig. 4.4B). The plunge of the cut-off line can be calculated if required (see Sections 3.10 and 3.11).

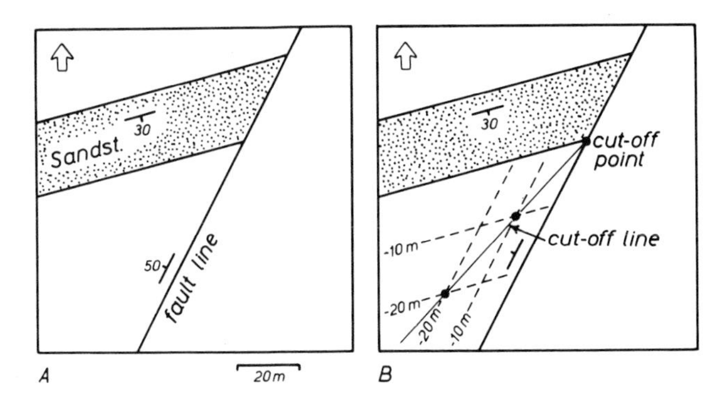

FIG. 4.4

In the field, the trend of faults is often easier to ascertain than the dip of a fault plane. Faults, being movement zones, often function as lines of erosional weakness which give rise to linear topographic features. The fault planes themselves are rarely exposed. The dip can sometimes be deduced from the form of its line of outcrop across irregular topography.

Their attitude relative to the structure of the country rocks allows faults to be classified. A *strike fault* has a strike parallel to the strike of the beds it displaces. A *dip fault* strikes parallel to the dip direction of the surrounding beds. A *bedding plane fault* is parallel to the bedding (Fig. 4.5) and is therefore a variety of strike fault.

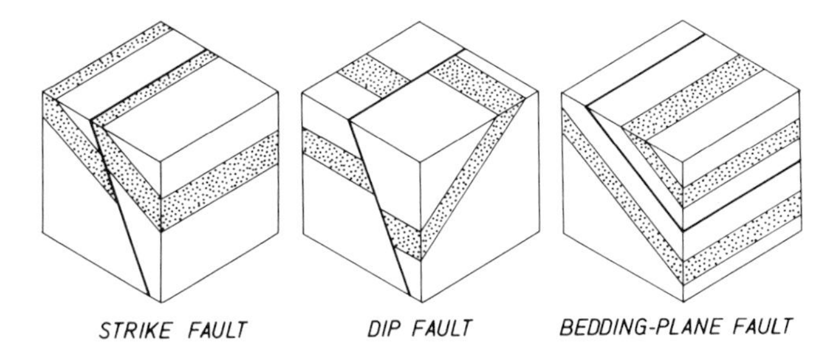

FIG. 4.5

4.2 Slip and separation

The *slip* of a fault is a measure of its actual displacement. The slip is the distance between two *points* on the fault plane, one on the foot wall and one on the hanging wall, which were originally coincident. Slip is determined by measuring the distance between equivalent recognizable points across the fault plane.

Separation has to do with the offset or shift of *beds or other geological surfaces* as observed on a specified outcrop or section surface. Separation is measured from displaced planes. In general, there is no simple agreement between the amount of separation and slip shown by a fault.

4.3 Separation terms

Separation is concerned with the distance between a planar surface in the hanging wall and its counterpart in the foot wall. This distance can be measured in various ways.

If the fault does not involve a rotation of one block relative to the other, the two displaced half planes will be approximately parallel. The shortest distance between two parallel planes is the distance measured in a direction which is perpendicular to these planes. This is the *stratigraphic separation* (*SS* in Fig. 4.6). It is the offset of beds across the fault expressed as a stratigraphic thickness.

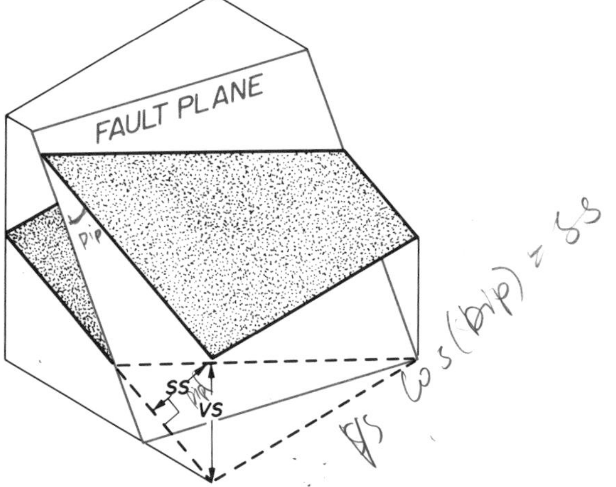

FIG. 4.6 Stratigraphic separation (**SS**) and vertical separation (**VS**).

If the separation is measured vertically, the distance is called the *vertical separation (VS* in Figs 4.6 and 4.7). Vertical separation is readily calculated from structure contour maps, being the difference in height between the displaced geological surface in the hanging and foot walls. The stratigraphic separation can be calculated from vertical separation if the dip of the beds is known, using

$$\text{Stratigraphic separation} = \text{vertical separation} \times \text{cosine (dip of beds)}$$

It follows from this that the stratigraphic separation equals the vertical separation in the case of horizontal beds.

Other measures of separation are distances measured in special directions with respect to the fault plane. The *strike separation* is measured parallel to the strike of the fault plane (Fig. 4.7). The offset of beds across a fault visible on a map of level terrain is equal to the strike separation. Strike separation can either be dextral (right lateral) or sinistral (left lateral) (see Fig. 4.8).

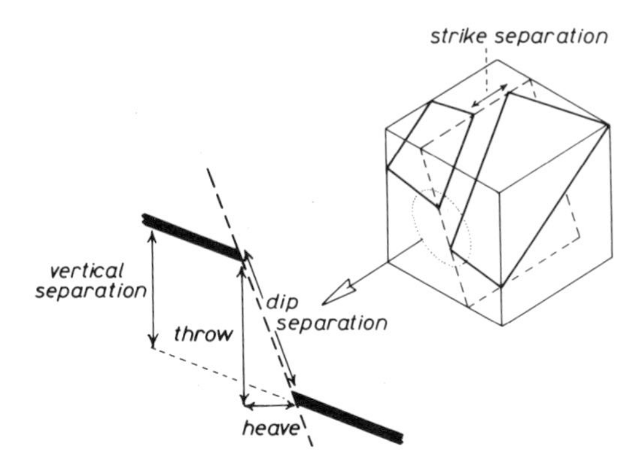

FIG. 4.7 Separation terms.

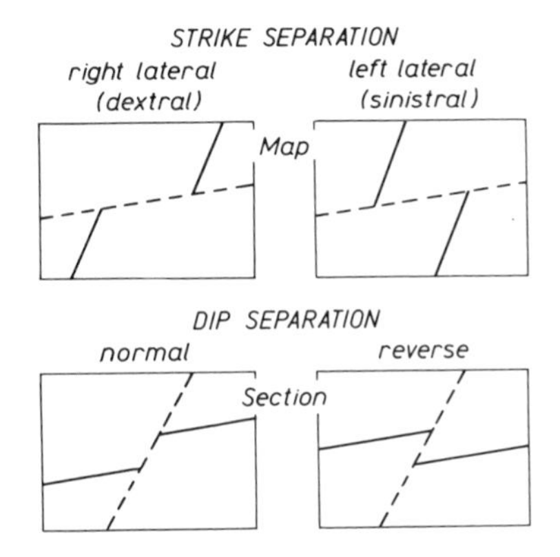

FIG. 4.8 Terms used to describe the sense of separation.

'Right lateral' means that the beds on the opposite side of the fault are offset to the right relative to the beds on the observer's side. *Dip separation* is the bed offset in the down dip direction of the fault (Fig. 4.7). The vertical component of the dip separation is called the *throw* and its horizontal component is the *heave*; both of these quantities can be seen in a plane perpendicular to the fault's strike. For a given fault, the ratio of the heave to the throw depends on the fault plane's dip; in this manner

tan (dip of fault) = throw/heave (Fig. 4.7)

where throw = dip separation × sin (dip of fault). Vertical faults have a throw but no heave. Based on the sense of dip separation on a fault, the block within which the beds show an apparent downward offset is referred to as the *downthrow side* of the fault. The downthrow side of a fault can also be easily determined from faults displayed on maps. This is done by oblique viewing of the map so that the observer looks down the dip of the displaced beds. By doing this, an impression is gained of how the beds would appear in cross-section. As an exercise, obliquely view the map in Fig. 4.9A along a line of sight which plunges 30° southwards (i.e. down the dip of the mudstone bed). From this it can be readily seen that the west side of the fault is the downthrow side. The downthrow side is denoted on many maps by a dash on one side of the fault line (see list of map symbols).

The heave is a meaningful measure of separation in relation to maps, as the example below shows.

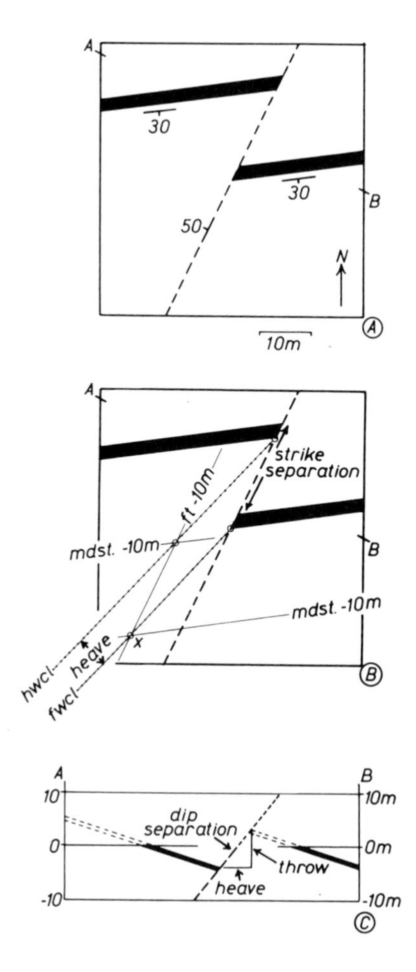

FIG. 4.9

Worked example

For the fault on the map (Fig. 4.9A) determine (a) the vertical separation, (b) the stratigraphic separation, (c) the strike separation, (d) the dip separation, and (e) the heave.

Structure contours and cut-off lines on the hanging (hwcl) and foot walls (fwcl) are constructed (using the method of the previous worked example).

The vertical separation is determined by selecting any point on the map and calculating the height difference of the two displaced parts of geological surface. For example at point x (Fig. 4.9B),

height of mudstone bed in hanging wall = −10 m
height of mudstone bed in foot wall = −20 m
height difference = −10 − −20 = 10 m = vertical separation
the stratigraphic separation = vertical separation × cos (dip) = 10 m × 0.64 = 6.4 m.

The strike separation can be measured directly from the map in this case (Fig. 4.9B) since the map is of an area of flat topography. Otherwise, this separation could be measured from the map in a direction parallel to the strike of the fault between the structure contours of the same height for the two sides of the shifted surface. The measured strike separation is about 20 m and the sense of offset is dextral. The dip separation and heave can be calculated in two ways. A vertical cross-section perpendicular to the fault's strike line can be drawn and these measurements made from it (Fig. 4.9C). Alternatively, the heave can be measured directly from the map (heave = 6 m) since it is the distance between the hanging wall and foot wall cut-off lines perpendicular to the strike of the fault. The heave can be used to calculate the dip separation, since

dip separation = heave / cos (dip of fault)
= 6 m / 0.64 = 9.4 m

The heave represents the gap between the cut-off lines, and as such indicates the width of ground not underlain by the mudstone bed.

The sense of dip separation is either normal or reverse. *Normal separation* involves a relative downward offset of the hanging wall. *Reverse separation* implies a relative upward offset of the beds in the hanging wall (Fig. 4.8). The fault in Fig. 4.9 shows normal separation of the mudstone bed.

4.4 Repetition and omission of strata

A straight line traverse through faulted strata will usually reveal that certain beds occur more than once *or* that certain beds do not occur at all. For any fault, beds can be either repeated or omitted depending on the direction of the line on which the occurrence of beds is recorded. The following worked example will allow the reader to formulate rules for the duplication and omission of strata resulting from faulting.

Worked example

The fault with dextral strike separation in the map in Fig. 4.10A displaces a sequence of strata number 1 to 14. Record the order of occurrence of strata in traverses *A-A′*, *B-B′*, *C-C′* etc., across the area. Which traverses involve repetition and which omission of strata? Formulate a general rule to determine which linear traverses show repetition and which omission.

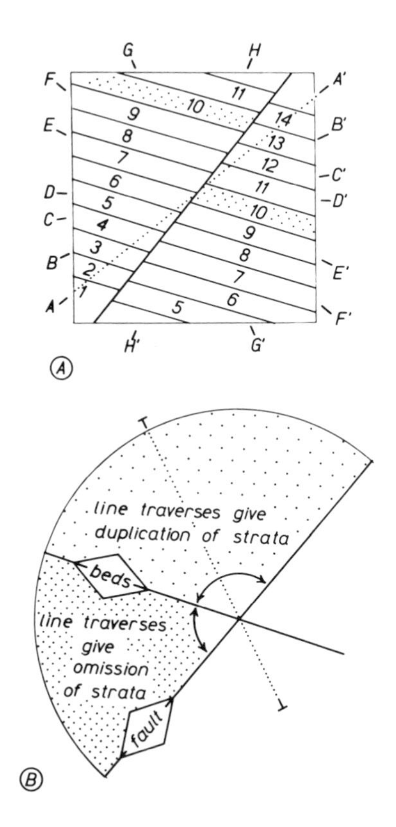

FIG. 4.10 Omission and repetition of strata resulting from faulting.

Line *A-A′* gives the sequence 1, 2, 3, 4, 5, 6, 10, 11, 12, 13, 14. This sequence involves omission of beds 7, 8 and 9. Lines *B-B′*, *C-C′*, *D-D′* also involve omissions of strata traverses; *E-E′*, *F-F′*, *G-G′* and *H-H′* involve duplication of beds. It appears, then, that the direction of the line traverse determines whether beds are cut out or duplicated. This exercise shows that the directions of the line traverses of each type fall into sectors. These two sectors (Fig. 4.10B) are bounded by the trace of the bedding and the trace of the fault plane. This rule can be extended to three dimensions.

4.5 Determining the slip of a fault

The separation or the offset of individual marker beds is not sufficient to deduce the actual movement or slip on a fault. To calculate slip it is necessary to recognize two points; one on the hanging wall and one on the foot wall, which were originally coincident. Two halves of a small object, say a pebble, fortuitously split by the fault, would provide a direct indication of the *net slip* (Fig. 4.11).

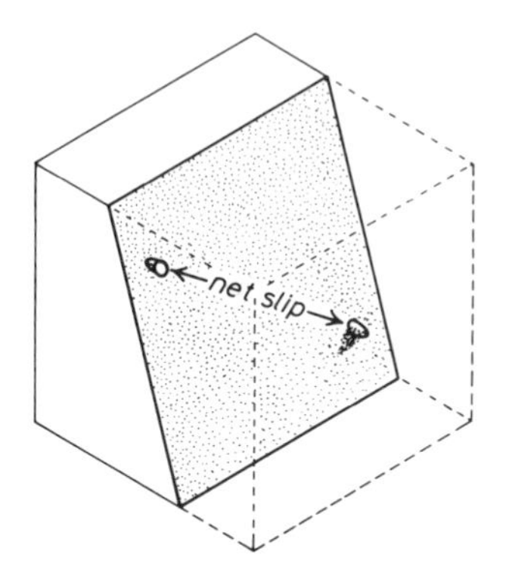

FIG. 4.11 Net slip.

Such displaced point objects are rarely found. Slip is more often calculated from points resulting from linear features which intersect the fault plane. Figure 4.12 shows an example of this where a fold hinge line provides a point (a hinge point), where the hinge line meets the fault plane.

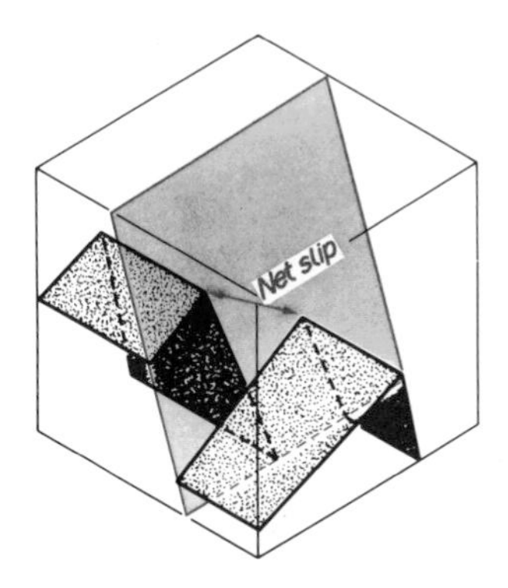

FIG. 4.12 Use of a displaced fold hinge line to calculate the net slip.

The *net slip* is the displacement of the two (hinge) points. Net slip has both direction and magnitude. The direction can be stated as the plunge of the line joining the two displaced points. The magnitude is the straight-line distance between the displaced points.

The points which are vital for the determination of slip are given whenever three planes intersect. One of these planes is the fault plane itself, the other two are a pair of any non-parallel planes. Such planes include fold limbs (as in the example above and Fig. 4.14), older faults, unconformities (explained in Chapter 5), sheet intrusions (Chapter 6), etc.

Worked example

Determine the net slip for the fault shown on the map in Fig. 4.13A. The fault displaces two planar calcite veins.

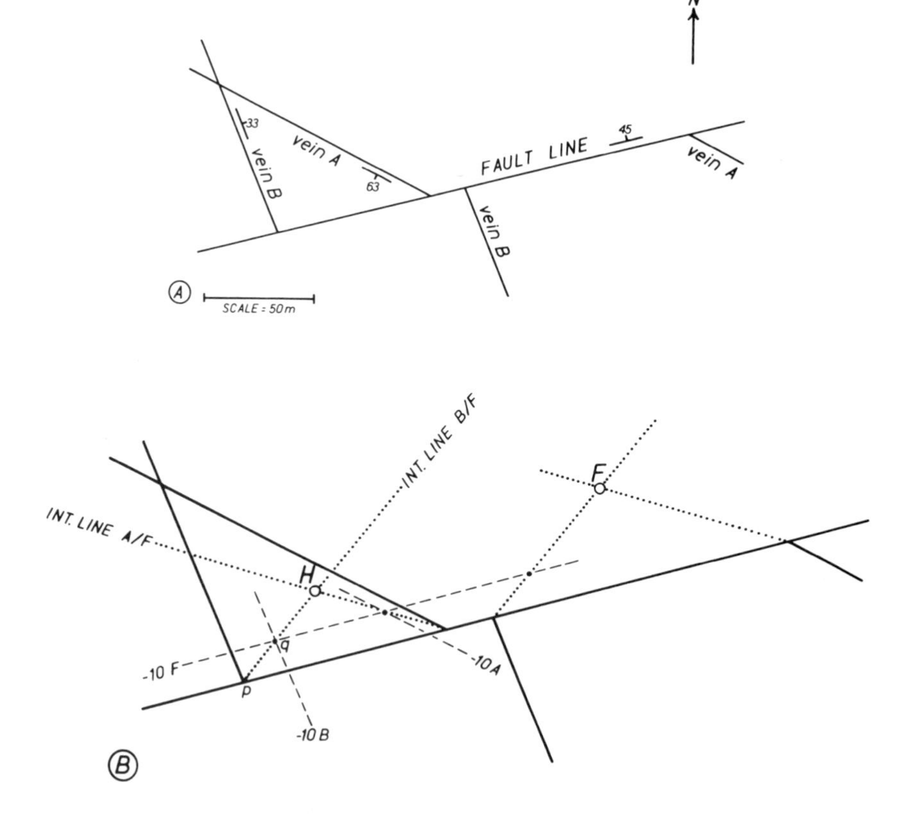

FIG. 4.13 Calculating the net slip of a fault.

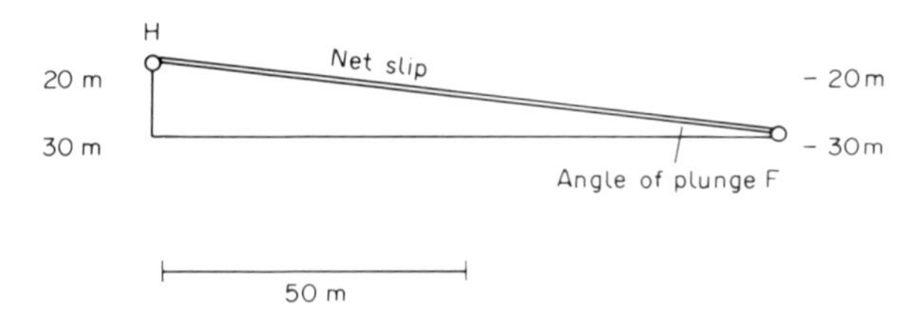

FIG. 4.13 **C**

For each of the three planes, one structure contour is drawn for a convenient height, say 10 m above the level of the ground surface. Using the usual method, lines of intersection are drawn for each pair of planes. These three lines of intersection cross at the point of intersection of the three planes (Fig. 4.13B). The three planes on the hanging wall intersect at *H*, on the foot wall at *F*. The line *FH* plunges in the direction *H*⟶ *F* (= 070°N) because point *F* is lower than *H*. From the difference in height and the horizontal distance *F-H* a triangle can be constructed which refers to a vertical plane through *F* and *H* (Fig. 4.13C). From this triangle the net slip (*FH*) and the angle of plunge can be read off. Therefore the plunge of the slip direction = 070-7 and the magnitude of net slip = 100 m.

FIG. 4.14 A fault which displaces two planes of different orientation (the limbs of a fold). Study the offset of the limbs. Which component of slip is more important, the dip-slip or the strike-slip component? (The outcrop surface is horizontal.)

4.6 Components of slip

The line representing the amount and direction of net slip of a fault can be resolved into components in various directions (Fig. 4.15). The *strike-slip component* is the component of the net slip in the direction of the strike of the fault plane. This component is either dextral or sinistral. The *dip-slip component* is the component of net slip in the fault plane's down dip direction. This component is either normal or reverse.

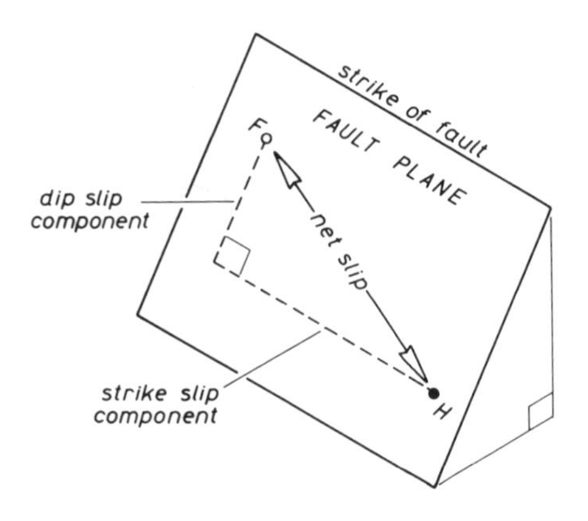

FIG. 4.15 Component of net slip.

Worked example

Calculate the strike-slip and dip-slip components for the fault in the map in Fig. 4.13A.

The strike-slip component is equal to length of *HF* projected on to the strike line of the fault (98 m in Fig. 4.13B). The sense of movement is sinistral. If the net slip and strike slip are known, the dip-slip component can be calculated because these three quantities define a right-angled triangle in the plane of the fault.

By Pythagoras,

$$(\text{net slip})^2 = (\text{strike slip})^2 + (\text{dip slip})^2$$

so that

$$\text{dip slip} = \sqrt{(\text{net slip})^2 - (\text{strike slip})^2} = \sqrt{100^2 - 98^2} = 20 \text{ m}.$$

The dip-slip sense is reverse (hanging wall is displaced upwards relative to the foot wall).

It is important to realize that there is no straightforward correspondence between slip and separation. For example, the dip-slip component will usually not be equal to the dip separation. To illustrate this consider Fig. 4.16, where a fault is observed to offset a bed. As no reference points can be recognized to allow the slip to be determined, a large number of possible slip directions could have given rise to the observed separation. Some of these possible slip directions are labelled 1 to 5 in Fig. 4.16. For instance, 5 is a potential slip direction which would have had a dextral strike-slip component, whereas the strike separation has a sinistral sense.

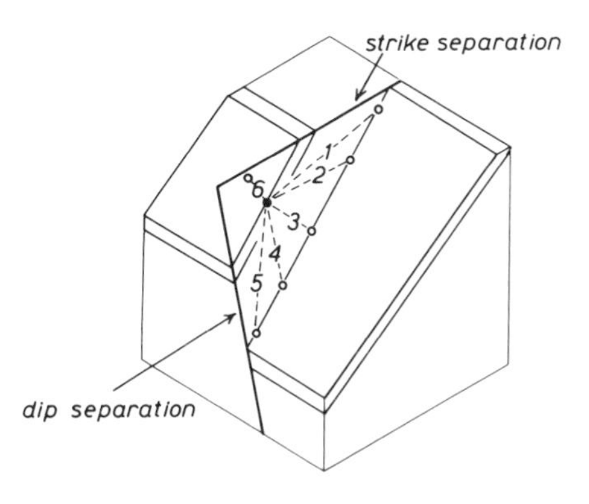

FIG. 4.16 To show that an observed separation shown by a planar marker bed could have been brought about by slip in a large number of directions within the fault plane.

Clearly the separation provides insufficient information for the complete determination of slip. Nevertheless, separation places certain limits on the possible slips. For instance, in Fig. 4.16 the slip labelled 6 with reverse dextral slip is clearly not a possible net slip for the fault. It can also be seen that line 3 is the shortest of all possible slips and therefore provides a minimum estimate of the slip magnitude. The equation below gives that estimate:

$$\text{minimum net slip magnitude} = 1 \Bigg/ \left[\frac{1}{(\text{dip separation})^2} + \frac{1}{(\text{strike separation})^2} \right]^{1/2}$$

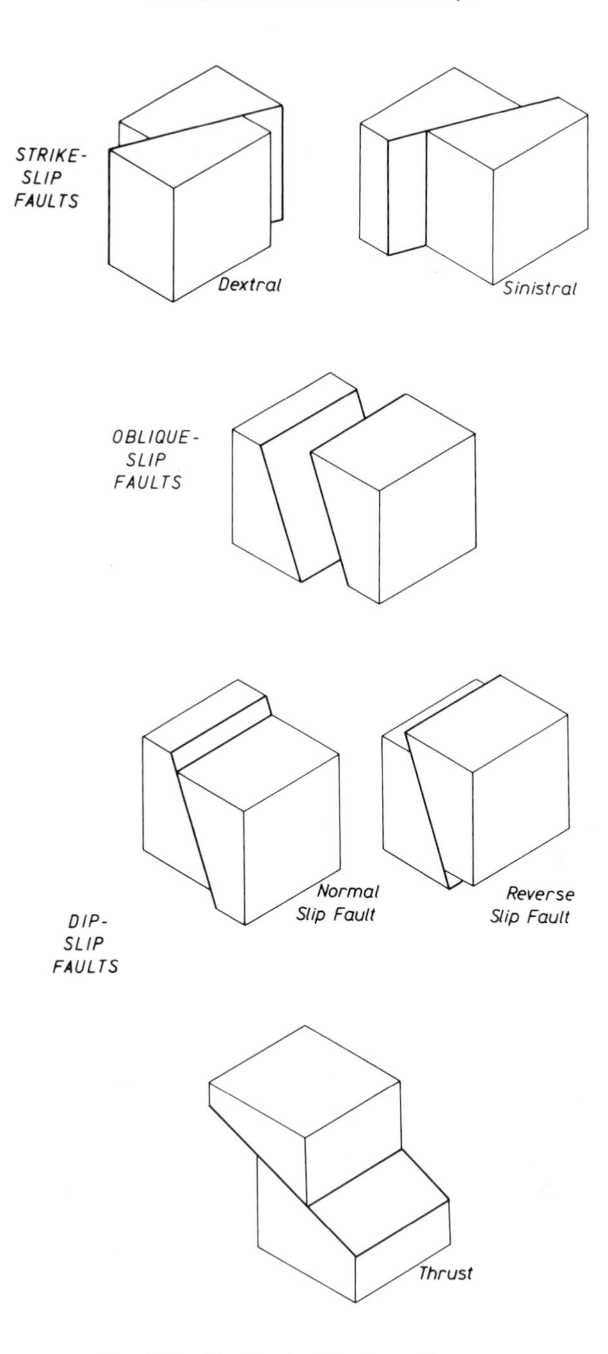

FIG. 4.17 The slip classification of faults.

4.7 Classification of faults based on slip

Table 4.2 below outlines the slip classification. These are illustrated in Fig. 4.17.

Table 4.2

Type of fault	Type of slip	Sub-types
Dip-slip fault	slip in down-dip direction; no strike component	normal-slip faults, reverse-slip faults
Strike-slip faults	slip parallel to strike of fault; no dip-slip component	dextral-slip sinistral-slip faults
Oblique slip faults	slip with dip slip and strike slip components	dextral normal-slip faults, sinistral reverse-slip faults etc.

Synonyms for strike-slip fault are *wrench*, *tear* and *transcurrent fault*. Normal-slip faults are commonly high-angle faults (Fig. 4.18A). High-angle reverse-slip faults are less common than normal-slip faults. Low-angle reverse-slip faults are called *thrusts* (Fig. 4.18B). Like strike-slip faults, thrusts can have net slips with magnitudes of several kilometres.

FIG. 4.18 Faults. **A**: Normal fault (or, more properly, a normal-separation fault), Bude, Cornwall.

FIG. 4.18 **B**: Thrust (a low-angled reverse fault), Torquay.

There are practical difficulties associated with the classification of faults. The slip classification is a logical scheme based on actual fault displacement but suffers from the disadvantage that the slip is only seldom determinable in practice. Separation, on the other hand, is readily measured but, as explained above, is only indirectly related to fault movement.

Whether a fault is described in terms of slip or in terms of separation it is advisable to name the fault in such a way as to make it clear which measurement has been made. For example terms like 'normal-separation fault' or 'normal-slip fault' should be used rather than simply 'normal fault'.

PROBLEM 4.1

The figure corresponds to a vertical cross-section showing the thicknesses and order of rock units encountered in four boreholes drilled along the line of section.

Examine the sequences in each borehole and compare these to the known stratigraphic sequence which is given in the column on the right.

In each borehole sequence, mark the position of fault or fold structures.

Interpret the structure of the cross-section by linking structures from adjacent boreholes.

PROBLEM 4.2

The aim of this exercise is to investigate the way fault separation relates to slip.

The map shows a vertical fault which cuts off a dipping sandstone bed.

The plane of the cross-section (lower diagram) coincides with the plane of the fault. The cross-section shows the trace, on the fault plane, of the sandstone bed in the block on the north side of the fault.

In the cross-section select any point on the base of the sandstone in the fault plane and, for an amount of slip given below, locate the position of its equivalent point in the fault plane on the south side of the fault. In the cross-section and map draw the displaced sandstone for the south side of the fault.

Measure the magnitude and sense of strike separation and dip separation for each of the following amounts of net slip:

	Strike-slip component	Dip-slip component
(i)	20 m (sinistral)	10 m (south side down)
(ii)	0 m	20 m (south side down)
(iii)	10 m (dextral)	30 m (south side down)
(iv)	30 m (dextral)	10 m (south side down)

Investigate the possibility that a fault may show neither strike nor dip separation.

PROBLEM 4.3

Determine the attitude (dip and dip direction) of the fault.

Draw a cross-section across the area in a direction at right angles to the strike of the fault plane.

Find the dip separation, throw, heave, vertical separation and strike separation of the fault.

Give a name to this fault.

PROBLEM 4.4

The map shows a fault (dashed line) which displaces a coal seam.

Determine the dip of the fault and the coal seam.

Construct the hanging wall and foot wall cut-off lines for the coal on the map and determine their plunge.

Draw a cross-section along a line at right angles to the strike of the fault.

Determine the strike separation, dip separation, throw, heave and the vertical separation of the fault.

Shade the parts of the area where coal can be encountered at depth. How is the heave of the fault important with respect to the areas underlain by coal?

PROBLEM 4.5

Beds dipping at 60° are displaced by two faults (f_1 and f_2); f_1 dips at 60° and f_2 is vertical. Which fault is the younger?

Determine the net slip of the younger fault.

Why is it not possible to find the slip on the older fault?

PROBLEM 4.6

The map shows Namurian rocks in West Lothian.

Study the map to identify fold hinges, fold axial traces and faults.

Determine the approximate plunge and plunge direction of the fold hinge lines.

Why does the strike separation vary along any one fault?

Estimate the net slip on the most northerly of the faults. (Assume that the faults are steep.)

PROBLEM 4.1

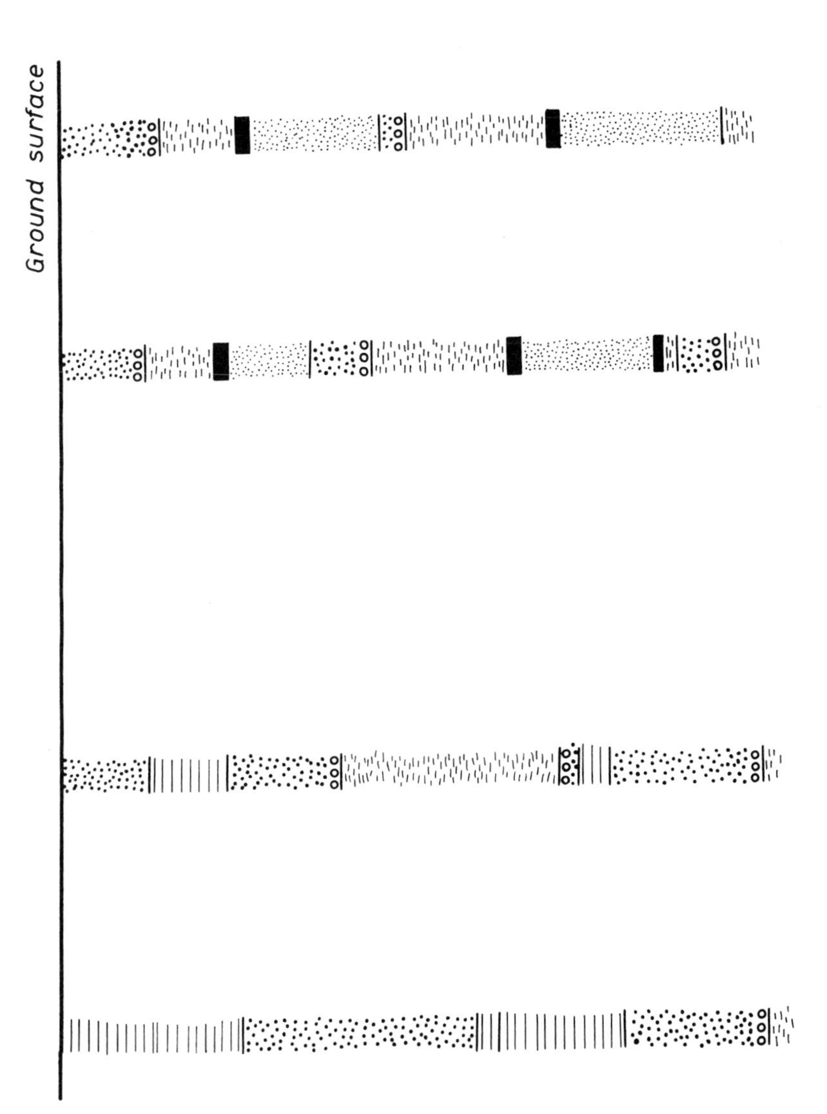

PROBLEM 4.2

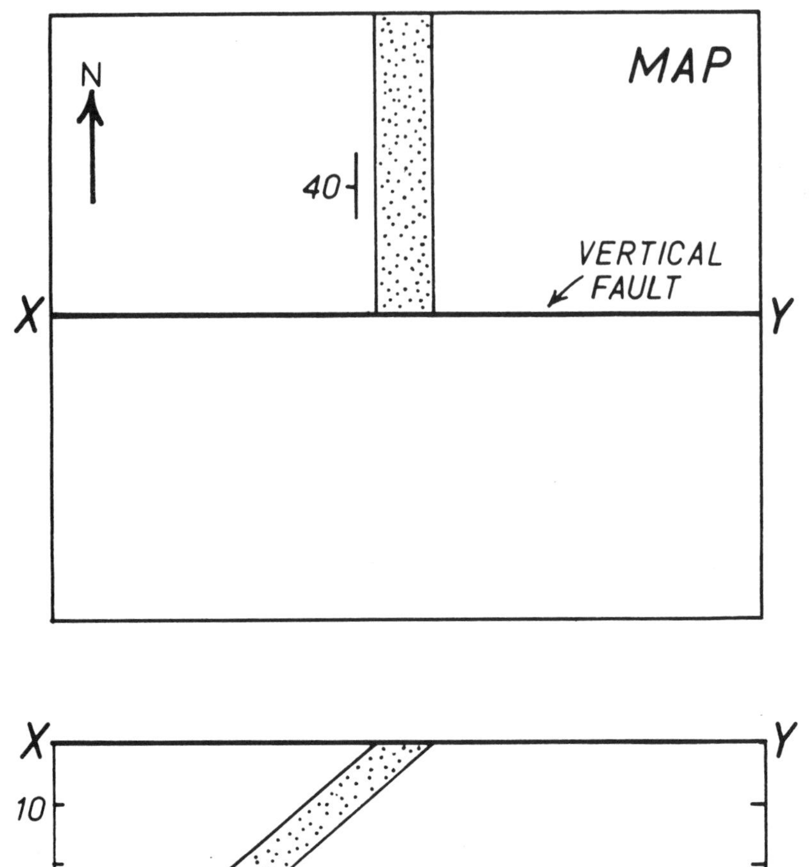
MAP
N
40
VERTICAL
FAULT
X
Y

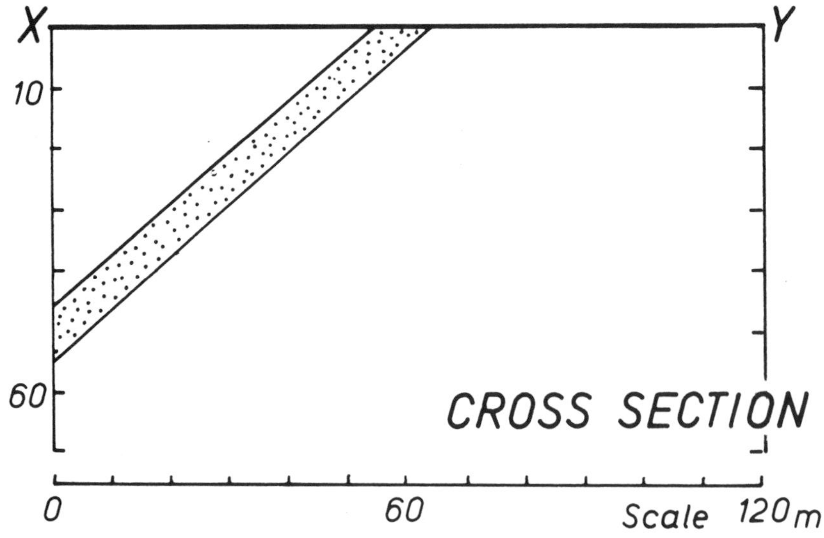
X
Y
10
60
CROSS SECTION
0
60
Scale 120m

PROBLEM 4.3

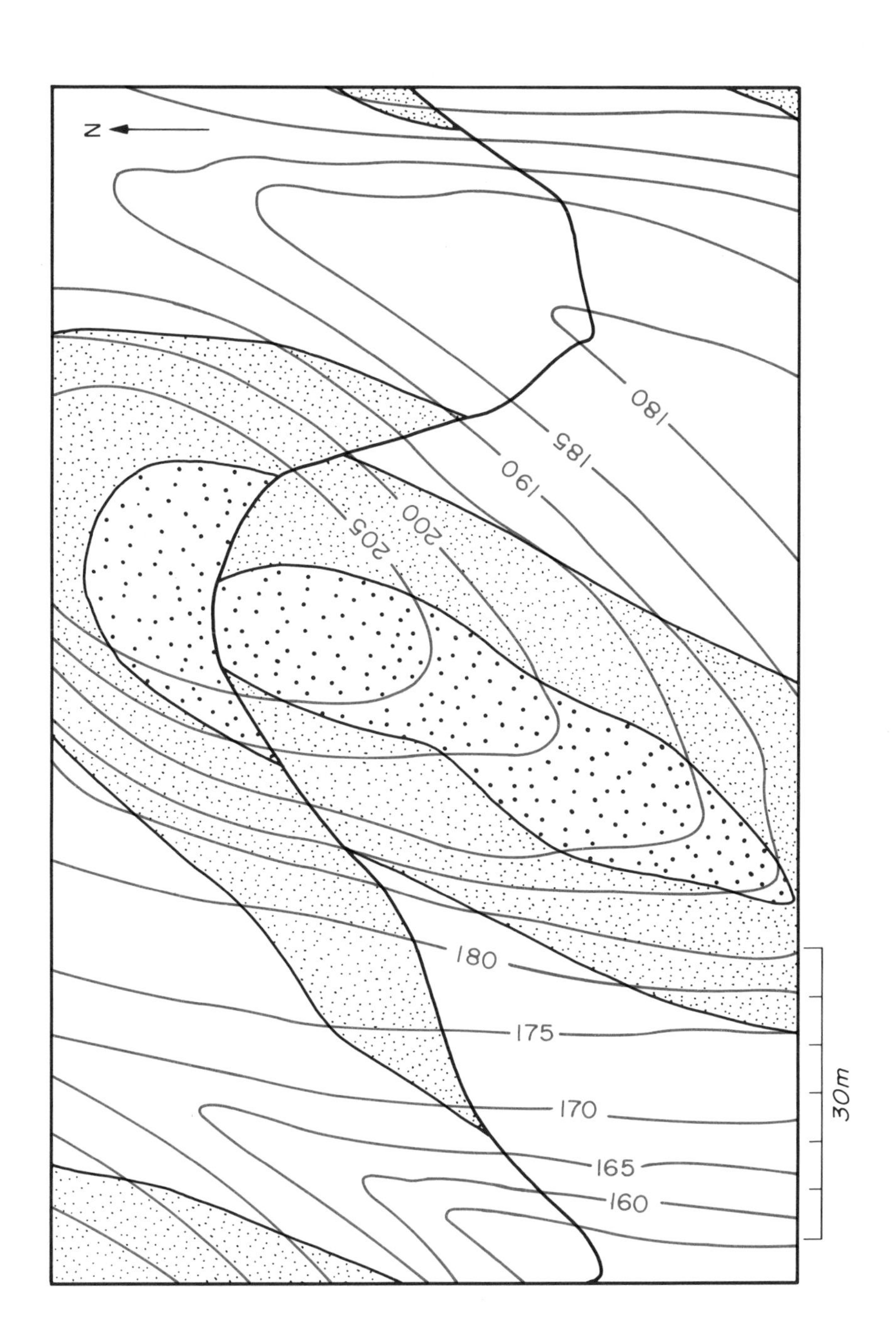

PROBLEM 4.4

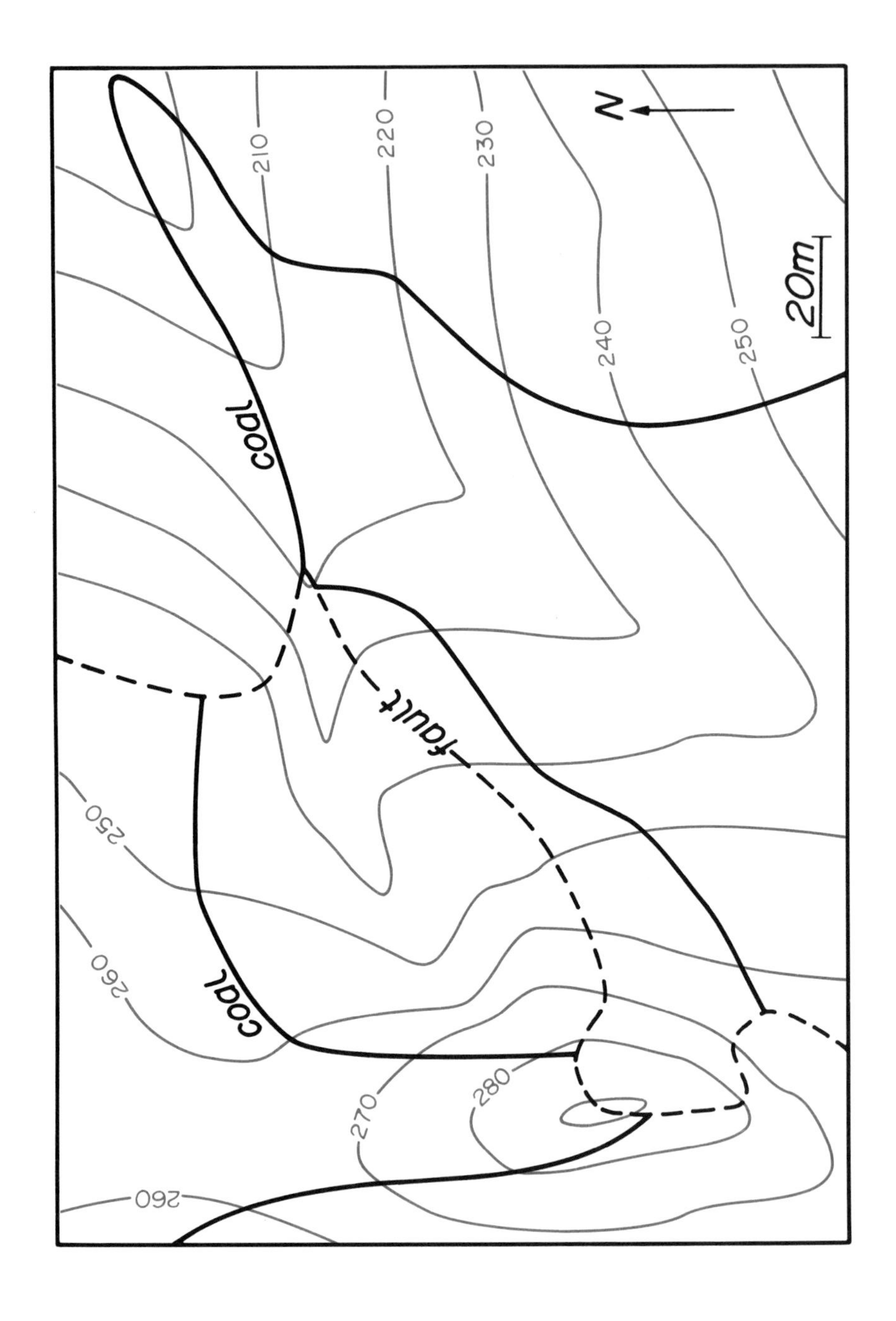

PROBLEM 4.5

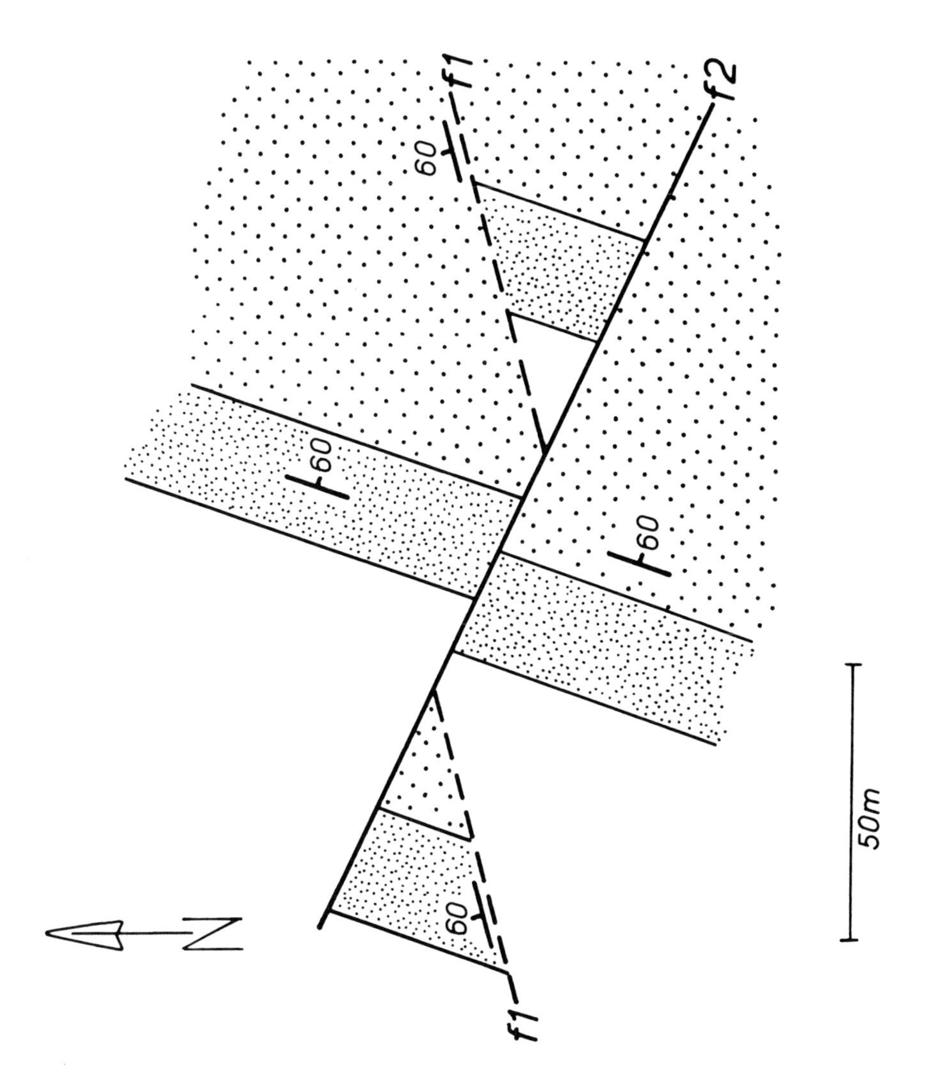

PROBLEM 4.6

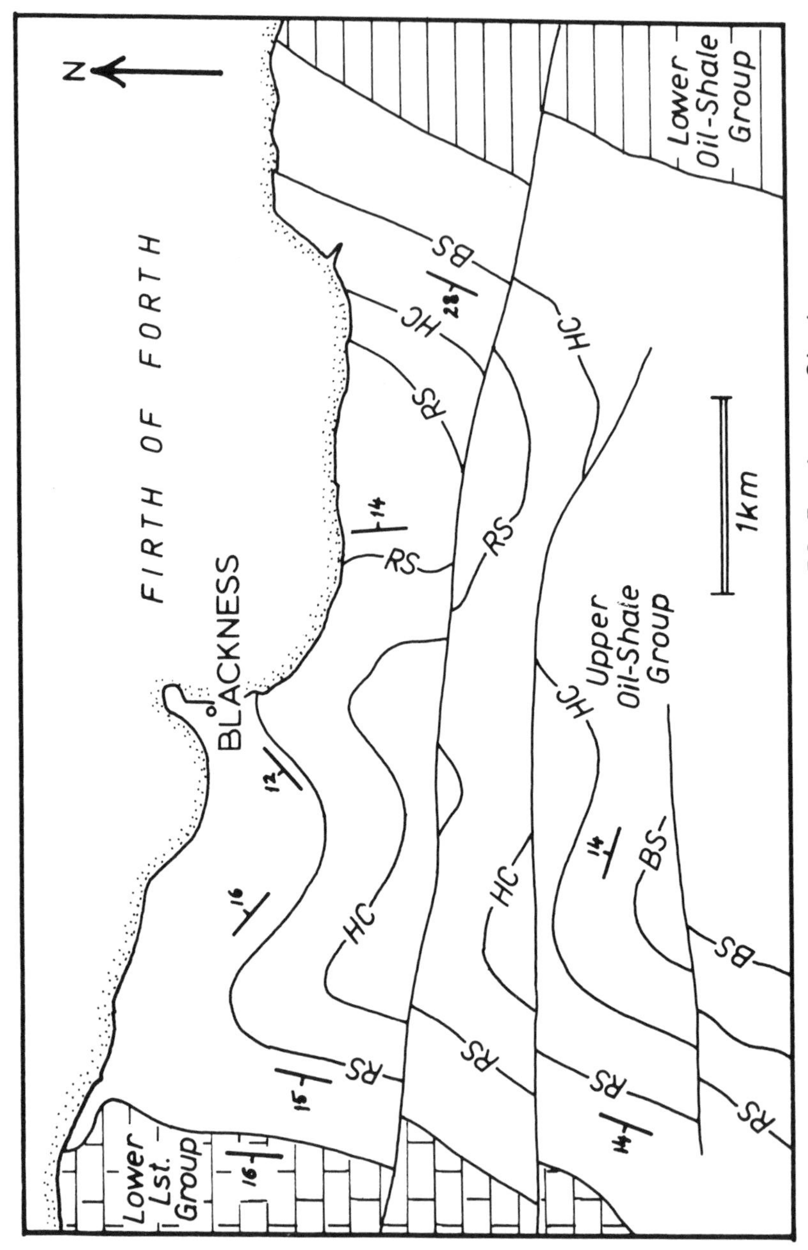
N
FIRTH OF FORTH
BLACKNESS
Lower Oil-Shale Group
Upper Oil-Shale Group
Lower Lst. Group
1km
BS
HC
RS
28
14
12
16
15
16
14
14
BS=Broxburn Shale, HC=Houston Coal, RS=Raeburn Shale

5

Unconformity

THE succession of sedimentary rocks present in any geographical area allows deductions to be made about the geological history of that area. For example, the nature of the sedimentary rocks indicates the physical environment in which the sediments were laid down and how this environment changed with time. In a particular area the sedimentary rocks themselves contain only a partial record of geological time since sedimentation is unlikely to have gone on uninterrupted. Periods of sediment deposition are likely to have alternated with intervals of non-deposition. Although the breaks in sedimentation do not leave behind rock products as hard evidence of their former existence, they do give rise to characteristic structural relationships between the rocks deposited before, and those deposited after, the interval. The relationship which bears witness to a period of non-deposition is called an *unconformity*. Because of a discontinuity in the sedimentation, later sediments come to rest *unconformably* on older rocks. The contact between the underlying rocks and the rocks which unconformably overlie them is termed the *surface of unconformity* or plane of unconformity. This surface may represent a considerable length of time, and could therefore be of great stratigraphic significance.

5.1 Types of unconformity

The exact nature of the relationship which defines an unconformity depends to a large extent on the geological events occurring during the period of non-deposition. If erosion takes place an irregular erosion surface is formed above the older rocks and defines the shape of the topographic surface upon which the younger group of rocks is deposited. This surface will become the surface of unconformity (Figure 5.1A). A *parallel unconformity* is where the beds above and below the surface of unconformity have the same attitude.

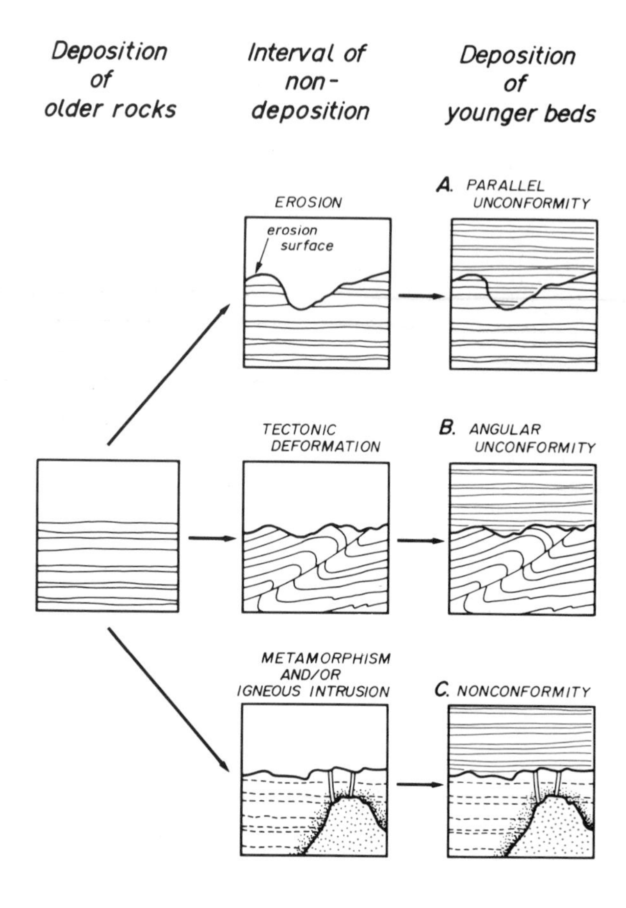

FIG. 5.1 Formation of unconformities.

With an *angular unconformity* (Fig. 5.1B) tilting or folding during the period of non-deposition gives rise to a misorientation of the rocks below the surface of unconformity relative to those above. When metamorphism and/or igneous intrusion takes place in the inter-depositional interval, the younger group of rocks rest in direct contact with metamorphic and/or igneous rocks. This type of unconformity is sometimes referred to as a *non-conformity* (Fig. 5.1C).

Figure 5.2 shows some natural examples of unconformities.

FIG. 5.2 Unconformities in the field. **A**: Portishead. **B**: Barry Island.

FIG. 5.2 **C**: Assynt.

5.2 Overstep and overlap

The most conspicuous feature of many unconformities is the way beds and, on a large scale, rock formations wedge out against the surface of unconformity (Fig. 5.2). On a small scale this same feature can be brought about by irregularities in the surface of unconformity, and can therefore occur in the case of parallel unconformities. On a regional scale, such wedging-out exhibited on maps by a tapering out of rock units against the surface of unconformity is likely to signify an angular unconformity. Overstep and overlap are terms which describe the angular relationship of stratigraphical boundaries to the surface of unconformity. *Overstep* applies to the sub-unconformity relationship where the surface of unconformity truncates stratigraphic boundaries (Fig. 5.3). With distance along the unconformity the overlying strata rest unconformably on successively older rocks and are said to 'overstep' them. Overstep usually owes its origin to tilting or folding during the pause in sediment accumulation. *Overlap* (sometimes called 'onlap') refers to the situation above the surface of unconformity. With overlap the sediments are deposited oblique to the surface of unconformity (Fig. 5.3). With distance along the unconformity, successively younger rocks rest on the unconformity plane. Successively younger rock units show greater lateral extent and thus 'overlap' the previously deposited units. This type of unconformity can result from deposition in a progressively expanding

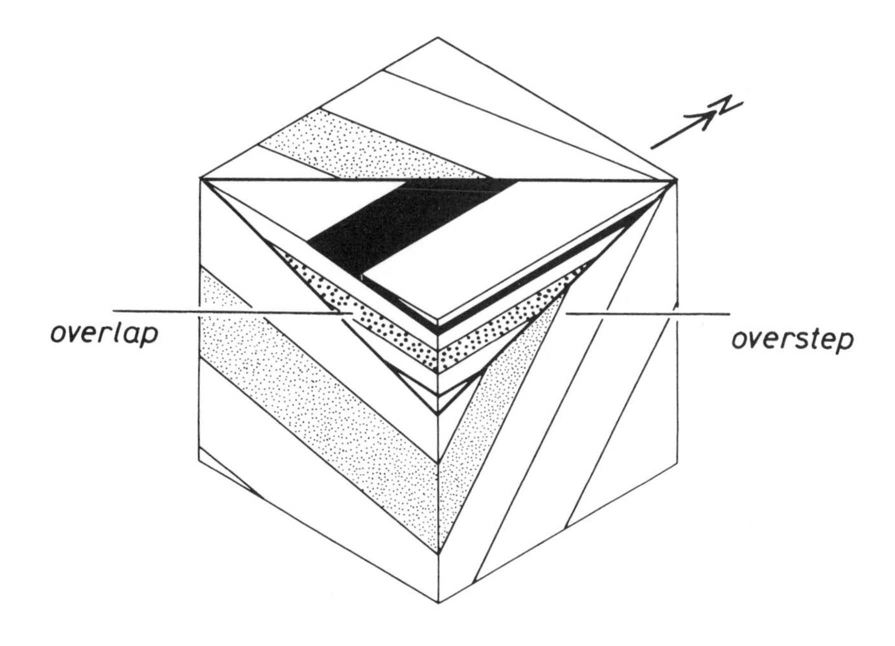

FIG. 5.3 Overlap and overstep.

sedimentary basin relating to crustal subsidence. Figure 5.4 shows examples of overstep recognizable from maps. Overlap and overstep are in no way mutually exclusive; unconformities may show both features.

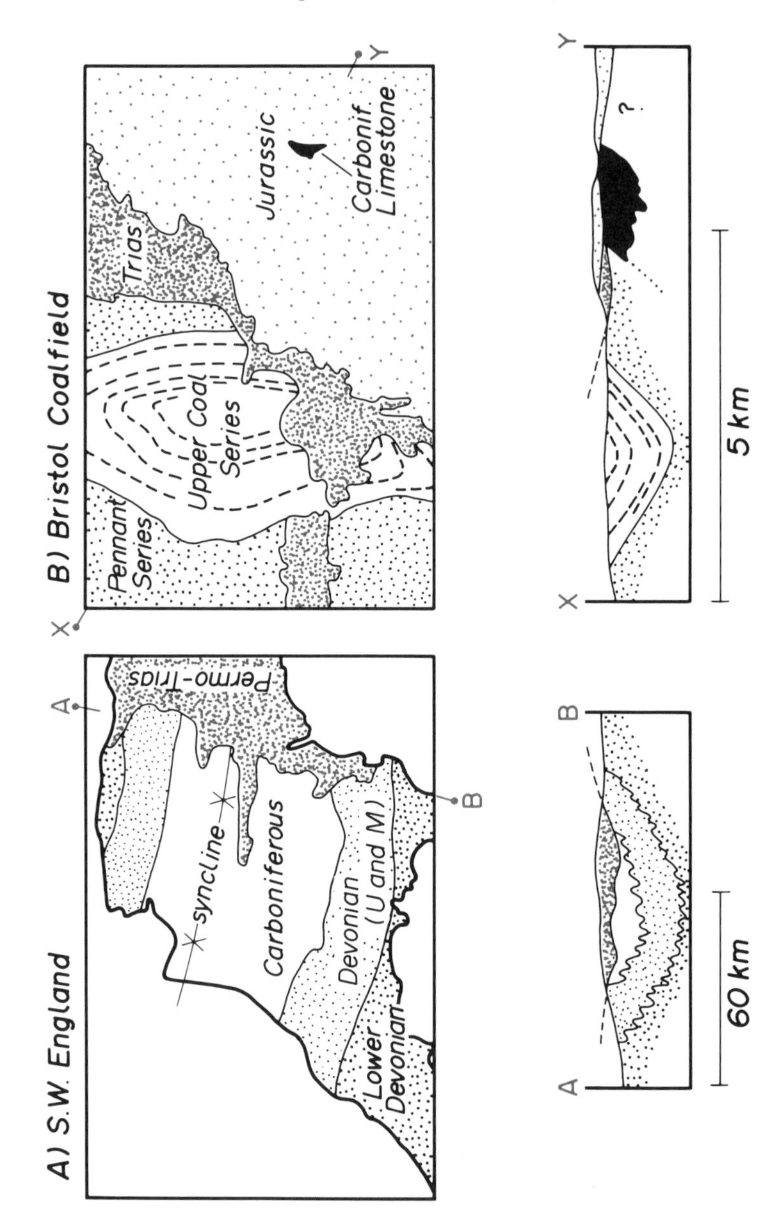

FIG. 5.4 Examples of unconformities displayed on maps and cross-sections.

5.3 Subcrop maps

A subcrop map (or palaeogeological map) represents the outcrop pattern of sub-unconformity rock formations on the surface of the unconformity. A subcrop map is how the geological map would look if the rocks which overlie

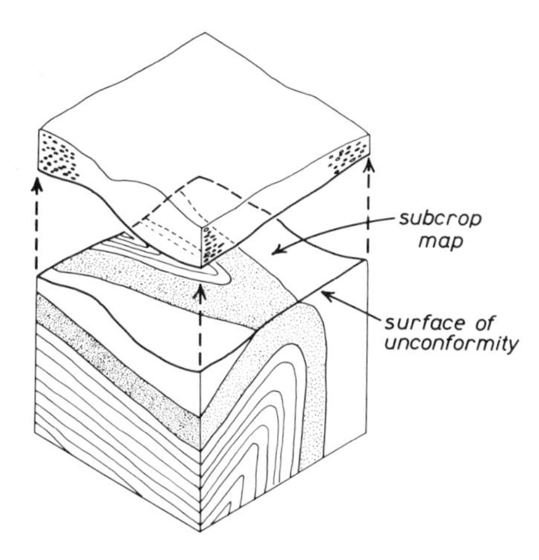

FIG. 5.5 The concept of the subcrop map.

the unconformity were to be stripped off (Fig. 5.5). Such a map has to be constructed from available data on the nature of the rock which immediately underlies the surface of unconformity. Data from boreholes which pass down through the unconformity provide a sound basis for the construction of a subcrop map. In some situations it may be possible to make predictions of the subcrop pattern from

(a) the rock types which are seen to underlie the unconformity at places where the plane of unconformity is exposed on the geological map;
(b) the known or assumed form of the surface of unconformity represented, for example, by structure contours; and
(c) the known or assumed attitudes of geological contacts of the formation which underlie the unconformity.

In practice, sensible predictions can be most easily made when the surface of unconformity and the contacts between formations below this surface are, or are assumed to be, planar.

Worked example

On the map (Fig. 5.6A) construct the subcrop of the thin coal seam on the pre-Permian unconformity surface.

The line of subcrop corresponds to the intersection of the coal seam with the surface of unconformity. If both surfaces are planar, the line of subcrop will be straight. This straight line will join all points where the coal crops out on the surface of the unconformity. It will therefore run as a straight line between points X and Y (Fig. 5.6B).

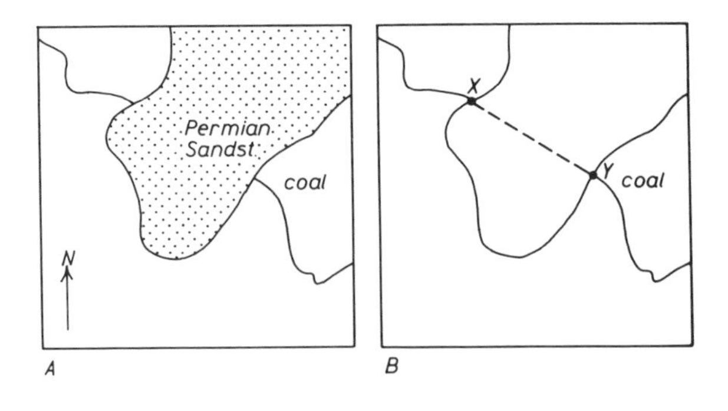

FIG. 5.6

Worked example

On the map (Fig. 5.7A) construct the subcrop of the thin coal seam.

From the given dips of the coal seam and surface of unconformity, calculate the trend of the line of their intersection using the method described in Section 3.11. Using the calculated trend, draw the subcrop of the coal through the point *Z* on the map (Fig. 5.7B). Vertically below all points on the map on the NW side of this line it is possible to encounter coal at depth; to the SE of this subcrop, the coal is absent.

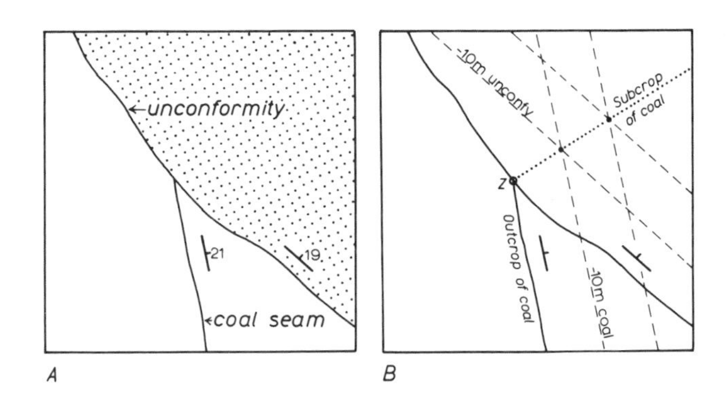

FIG. 5.7 Calculating the subcrop of a coal seam.

5.4 The geological usefulness of unconformities

Geological map interpretation has two main facets. The first is geometrical, and much of this book has been devoted to the techniques for deducing the form of structures from the patterns displayed on maps. The second aspect is historical in nature, and is concerned with making deductions about the relative ages of geological phenomena displayed on maps. Unconformity is important in this second respect, since it allows the ages of folding, faulting, metamorphism and igneous activity relative to that of sedimentation to be established.

As an example let us examine a map of the pre-Permian unconformity in SW England (Fig. 5.4A). Folds are present in the rocks below the unconformity and are absent in the rocks above. Folding must therefore have occurred in the time interval represented by the surface of unconformity. The youngest rocks below the unconformity involved in the folding are Upper Carboniferous (Westphalian) in age. The oldest rocks above the unconformity are Permian in age. Folding must therefore have taken place in the intervening time interval.

PROBLEM 5.1

Study the map and identify an unconformity. (The line labelled *f* is a fault line.)

What type of unconformity is it?

Draw a subcrop map of the formations which underlie the unconformity.

Discuss the age and movement on the fault.

PROBLEM 5.2

Determine the net slip of the fault. What name is given to this type of fault?

Identify an unconformity on the map and determine the dip of the surface of unconformity.

Why does the fault appear not to displace the unconformity?

Draw a subcrop map of the rocks below the unconformity.

PROBLEM 5.1

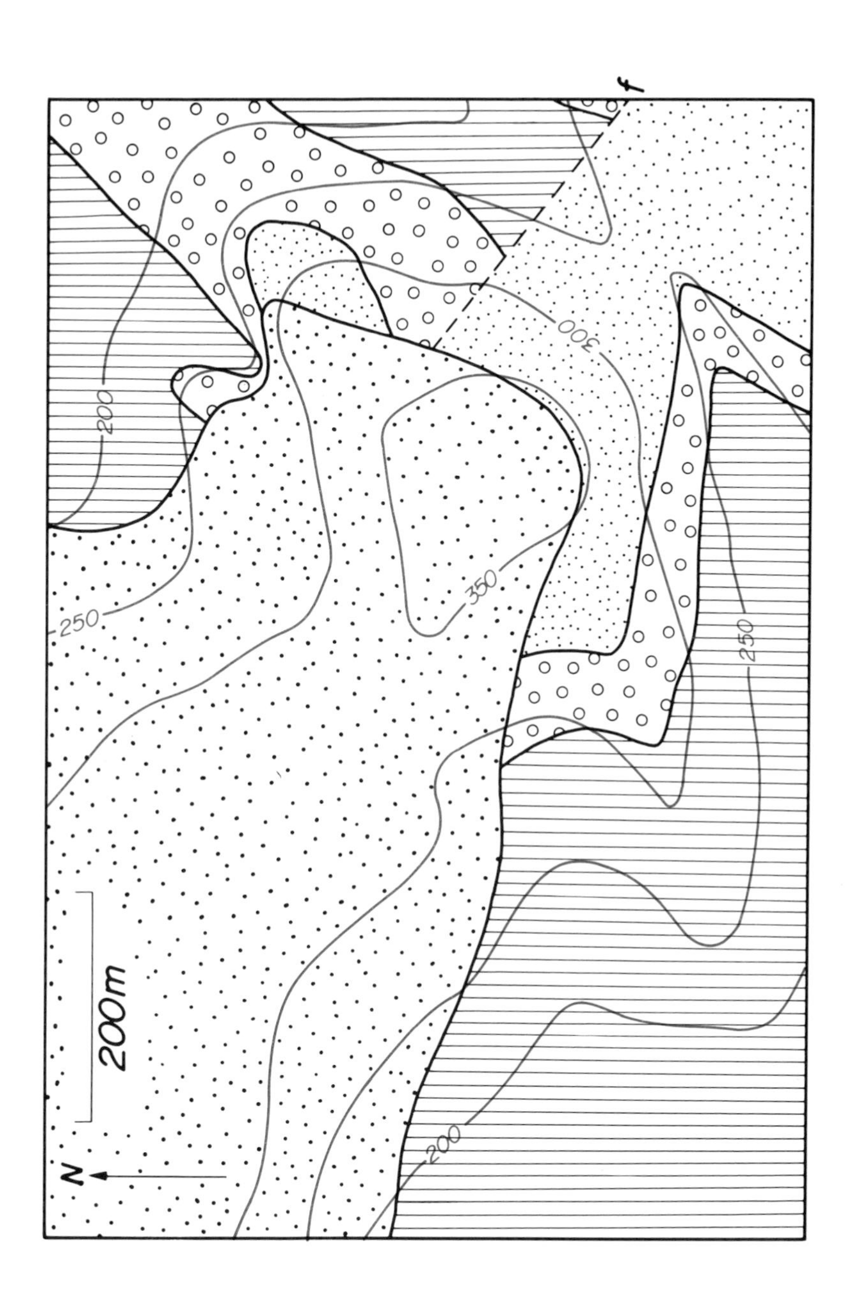

PROBLEM 5.2

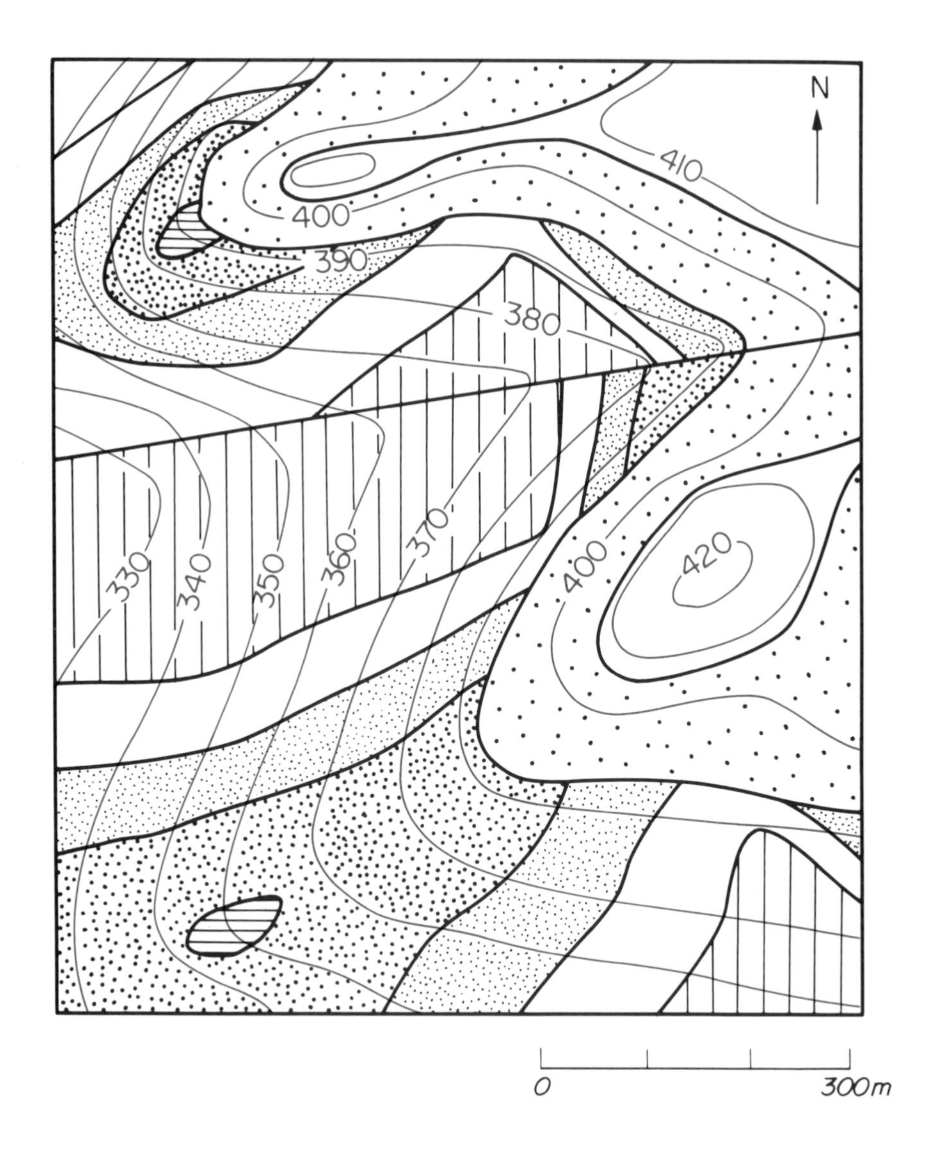

6

Igneous Rocks

THE maps so far discussed have involved layered piles of rocks characteristic of successions of sedimentary rocks and sequences of metamorphic rocks derived from such sediments. With the exception of unconformities, which represent sedimentary breaks or discontinuities, the layered structure of such rocks arises from the deposition of beds of sediment, one upon another.

A sequence of sediments may contain within it bodies of rocks belonging to an important group called *igneous rocks*. Igneous rocks are formed by the solidification of molten rock or magma. They are described separately because they may come into position in a way which is quite different from the orderly 'layer cake' manner typical of sedimentary rocks. For example they can be injected into a sequence of rocks like jam into a doughnut.

The discussion of igneous rock varieties (in terms of composition) is beyond the scope of this book. We will concentrate here on the classification of igneous rock bodies in terms of their form and structural relationship to adjacent rocks. These aspects are the most relevant to the appearance of igneous rocks on geological maps. This classification is set out in Table 6.1.

Table 6.1

		Examples
Intrusive igneous rocks	Concordant bodies	Sills
	Discordant bodies	Dykes, volcanic necks, batholiths
Extrusive igneous rocks		Lava flows, volcaniclastic deposits

6.1 Intrusive igneous rocks

These are bodies formed by the injection of magma into an existing sequence of rocks. Many have a tabular form and are called *sheet intrusions*. *Sills* have been intruded along the layering or bedding of the sequence and are examples of concordant sheet intrusions (Figs 6.1A and 6.3). When displayed on a map, a sill will follow the trend of other geological units (Fig. 6.3). The Whin Sill of northern England can be traced on the map from the

Farne Islands, off the Northumberland coast, southwards for some 170 km to Teesdale. Its average thickness is about 30 m and it is intruded into Carboniferous rocks but not always along exactly the same stratigraphic horizon. The Whin Sill is concordant over much of its trace but locally discordant. A sill which is locally cross-cutting is called a *transgressive sill* (Fig. 6.1B).

FIG. 6.1 Concordant sheet intrusions. **A**: Intrusion within deformed Precambrian quartz schists (Dombas, Oppdal, Norway). **B**: Transgressive sill (Banks Island, Northwest Territories, Canada). (*Geological Society of Canada.*).

Dykes are sheet intrusions which are discordant, i.e. they cut across the layering of the rocks they are intruded into (Figs 6.2 and 6.3). They usually have steeply dipping contacts. Many dykes occur in the Inner Hebrides region, which was a centre of intrusive igneous activity during the Tertiary. These dykes form swarms which are sets of parallel or radiating dykes. As can be seen from a geological map of the southern part of the Isle of Arran, dolerite dykes belonging to the swarm can be very closely spaced. On a map dykes stand out because of their discordant nature. Their relatively straight course relative to the other formations on a map is explained by their near-vertical attitude. Dykes can often be followed in the field because of their effect on the topography. In spite of their name, dykes do not always form a positive feature (such as a ridge) on the ground. Sometimes they weather out more than the rocks they are intruded into, and form a linear depression.

FIG. 6.2 Dyke of basic igneous rock (dolerite) cutting Lewisian gneiss with banding that dips gently to the right of the picture. Lewis, Outer Herbrides.

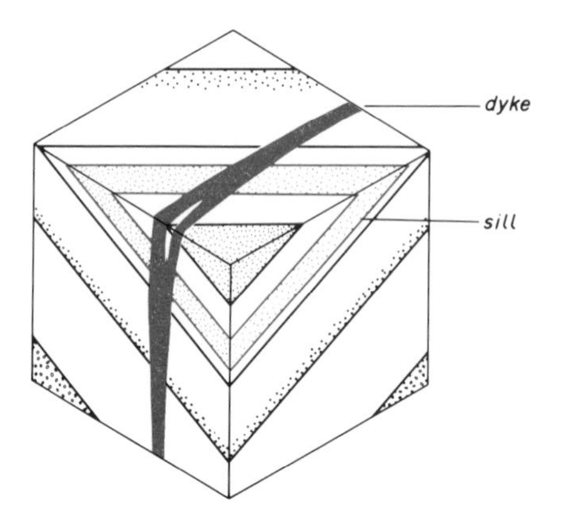

FIG. 6.3 Sheet intrusions.

Not all intrusive igneous bodies are sheet-like in form. *Batholiths* are large bodies usually composed of acid plutonic rock. Their contacts may be irregular in shape but the body often has an overall dome-like form with an upper contact (roof) which is more flat-lying than the steeper inclined walls (Fig. 6.4). In detail the upper part of the body may protrude into the *country rocks* of the roof to form a *cupola*. If the roof has a depression in it, the country rock will extend to a deeper level into the batholith. The resulting structure is a *roof pendant*. By erosion removing the roof of country rock, the igneous rocks making up the batholith may crop out at the surface. They are typically bodies which are roughly circular or oval in plan. Depending on the exact level of the erosion surface relative to the height of the roof, roof pendants and cupolas may form isolated closed contacts between igneous rock and country rock on the map. *Stocks* are smaller versions of batholiths. In SW England six granite stocks which occur at Dartmoor, Bodmin Moor, St Austell, Carnmenellis, Land's End and the Scilly Isles may represent cupolas of a single larger batholith at depth (Fig. 6.5). The country rocks adjacent to batholiths (and to a lesser extent some minor intrusions) often show marked evidence of being affected by the heat emanating from the intrusion. This metamorphism of rocks produced by their proximity to an intrusion is called *contact metamorphism*. The zone consisting of metamorphosed rocks is called a *metamorphic aureole*, and is displayed on maps as a zone encircling or running parallel to the contacts of the intrusion.

Volcanic necks or *plugs* represent the solidified magma and other volcanic rocks filling the pipe (vent) of former volcanoes of central type. Erosion produces a roughly circular map pattern which is a cross-section of the cylindrical pipe.

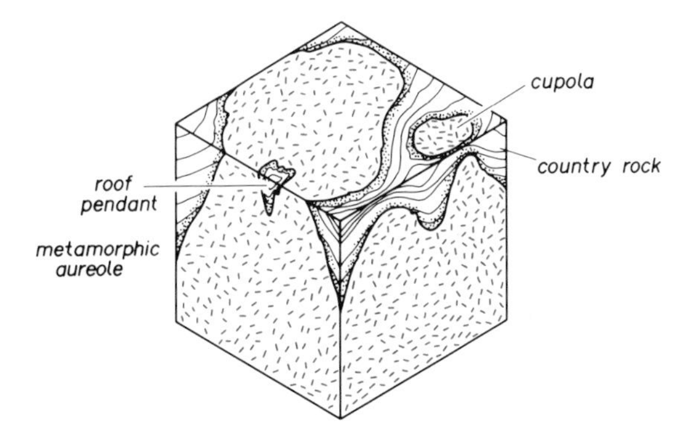

FIG. 6.4 Features associated with batholithic intrusions.

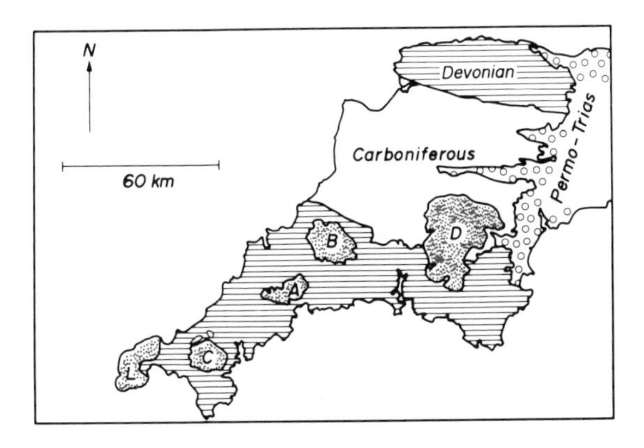

FIG. 6.5 The SW England batholith.

The subsurface form of discordant intrusions is frequently difficult to deduce from the form displayed on a map. For example, Figure 6.6A shows two isolated outcrops of igneous rock. Are they truly separate bodies or are they really connected at depth? When the contact is cut by topographic contours on the map, structure contours can be constructed and the dip deduced in the usual way (Section 2.9). However, the topographic relief is often strongly influenced by the presence of a body of igneous rock which denudes differently from the country rock. For example, a stock could form an area of upland with topographic contours following the contact of the intrusion (Fig. 6.6B). In these situations the structure contour method should be applied with caution. For example, the intrusion contact in Fig. 6.6B could easily be misinterpreted as having a horizontal attitude from the fact that it runs parallel to topographic contours on the map (compare Section 2.14).

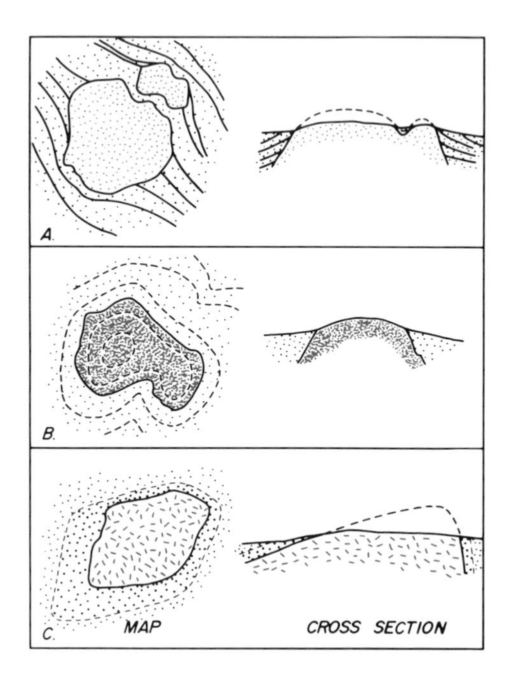

FIG. 6.6 Relations between the intrusion and structures in the country rock.

The uneven width on the map of a metamorphic aureole may provide a hint as to the variable dip of an intrusion's contacts (Fig. 6.6C).

The discordant nature of a contact may not be obvious on a map (Fig. 6.7). It should be realized that certain sections through a discordance will show a concordant relationship. Do not base your interpretation solely on a relationship at one point of the map. The intrusive character of an igneous body may be made less obvious by later deformation of the rocks of the area. The strain (distortion) suffered by the rocks has the effect of modifying angles between planar structures. In this way an angle of discordance between a dyke and the enclosing rocks could be reduced. The result will be to make the dyke appear as sill. Intrusive igneous bodies can also provide vital evidence for establishing a geological history of an area. Figure 6.8 shows a dyke which intrudes

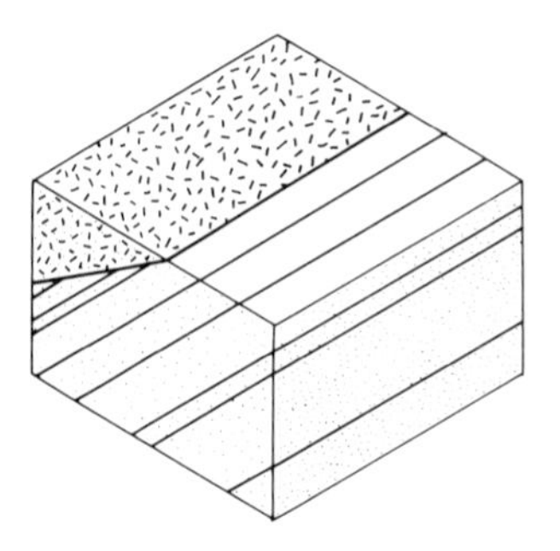

FIG. 6.7 Discordant intrusion can appear concordant on some cross sections.

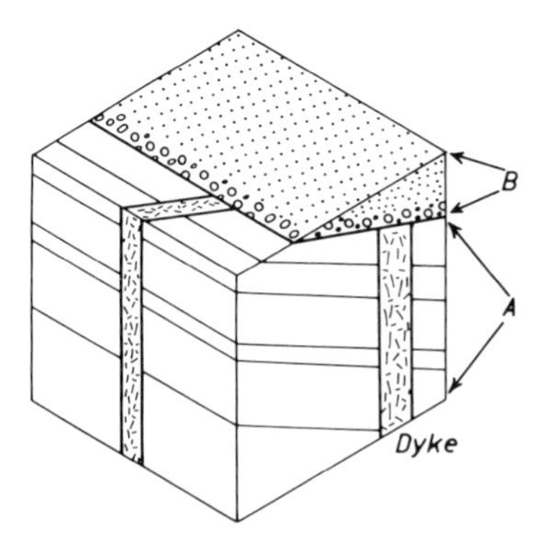

FIG. 6.8

a sequence of rock labelled *A*. The intrusion of the dyke clearly postdates the rocks of that sequence. The dyke does not intrude the sequence labelled *B*. We can conclude from this that sequence *B* was formed after dyke intrusion. The contact between *A* and *B* must represent the interval of time within which dyke intrusion occurred. This contact is therefore a plane of unconformity.

6.2 Extrusive igneous rocks

These rocks are formed from the products of volcanoes. Outpourings of lava on the earth's surface, and accumulations of fragments ejected from a volcano (*volcaniclastic rocks*), form sheets of volcanic rock. These rocks, extruded on land or under the sea, will rest on the rocks which form the subsurface at the time of eruption. They can lie comformably or unconformably with respect to the rocks beneath and above. When conformable, these sheets of volcanic material could resemble sills when displayed on the geological map. The transgressive or bifurcating (splitting) nature of sills will sometimes allow them to be distinguished from lava flows. The presence of contact metamorphic alteration of the overlying country rock would indicate a sill rather than an extrusive igneous rock. If the igneous rock includes fragments (*xenoliths*) of the overlying country rock this would also rule out an extrusive origin.

PROBLEM 6.1

Rock types 1, 2 and 3 are intrusive igneous rocks and the rock types labelled *m* are contact metamorphic rocks. List the types of intrusions present in the area. Faults are shown by a dotted line (4 in key). Give a name to the types of faults present and deduce as much as you can about the slip on these faults.

List, in order of time, the geological events which have affected the area.

PROBLEM 6.2

The map shows the geology of part of the Assynt District in Sutherland (NW Highlands of Scotland).

The rocks labelled *G* are Lewisian gneisses (coarse-grained, banded metamorphic rocks) of Precambrian age. The banding (layering) in the gneisses is strongly folded. Rock types *P* and *Q* are unmetamorphosed sediments, Cambrian in age. *S* are unmetamorphosed sandstones (Torridonian). *D* are igneous rocks.

Estimate the dip and dip direction of the Cambrian rocks.

Study the contact between the Lewisian gneisses and the Torridonian sandstones. For example, does the dip of this contact agree with the dips within the Torridonian Sandstone? Explain the shape of this contact and its significance.

List, in chronological order, the geological events which have occurred. What do the bodies of igneous rock tell us about relative ages of the rock units in the area?

Draw a cross-section along the line *A*-*B*.

PROBLEM 6.3

The geological map of New Cumnock (Dumfries) area (scale 1. 50,000, British Geological Survey, Sheet 15W) displays clearly the structural relationships of a variety of igneous rock bodies.

If this map can be consulted, deduce the geological history of that area.

PROBLEM 6.1

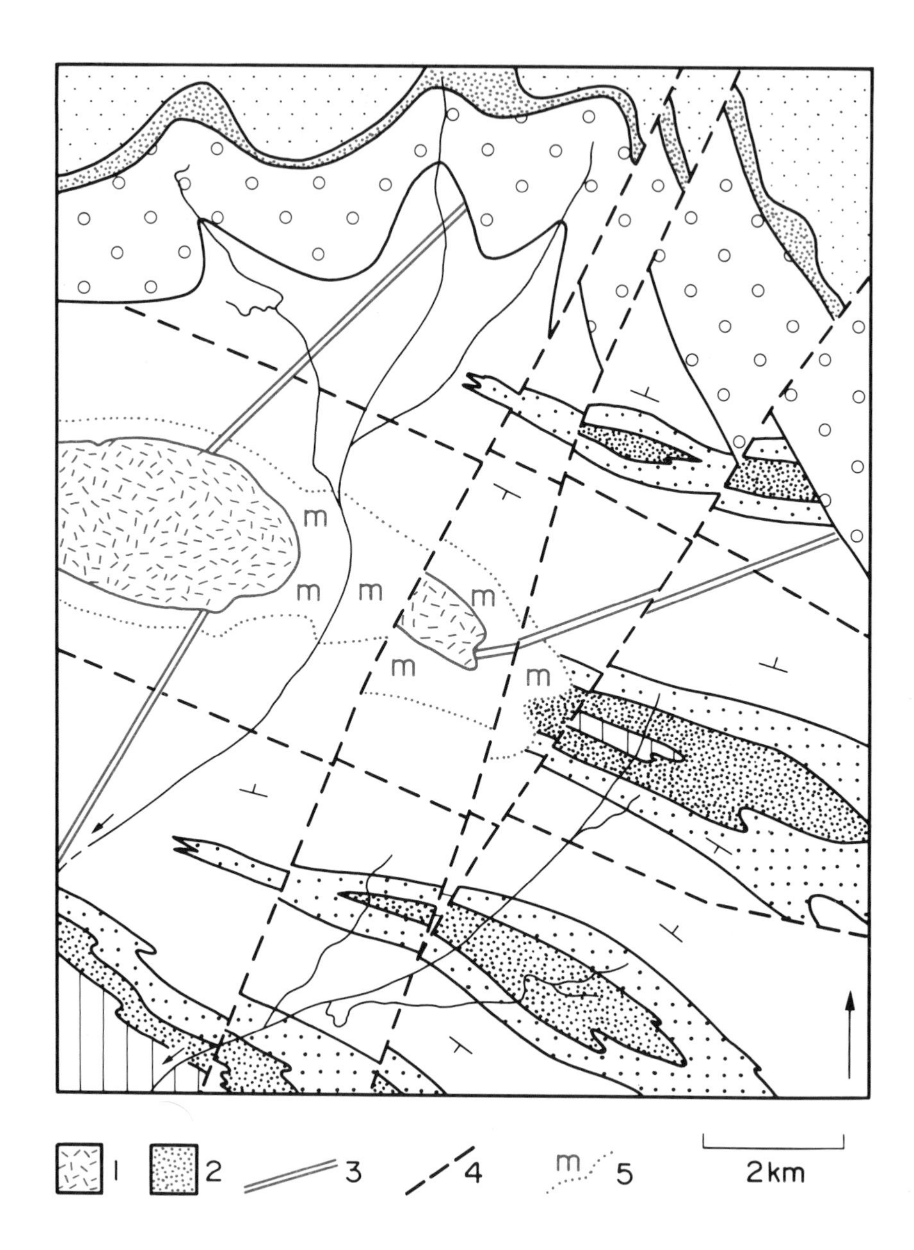

PROBLEM 6.2

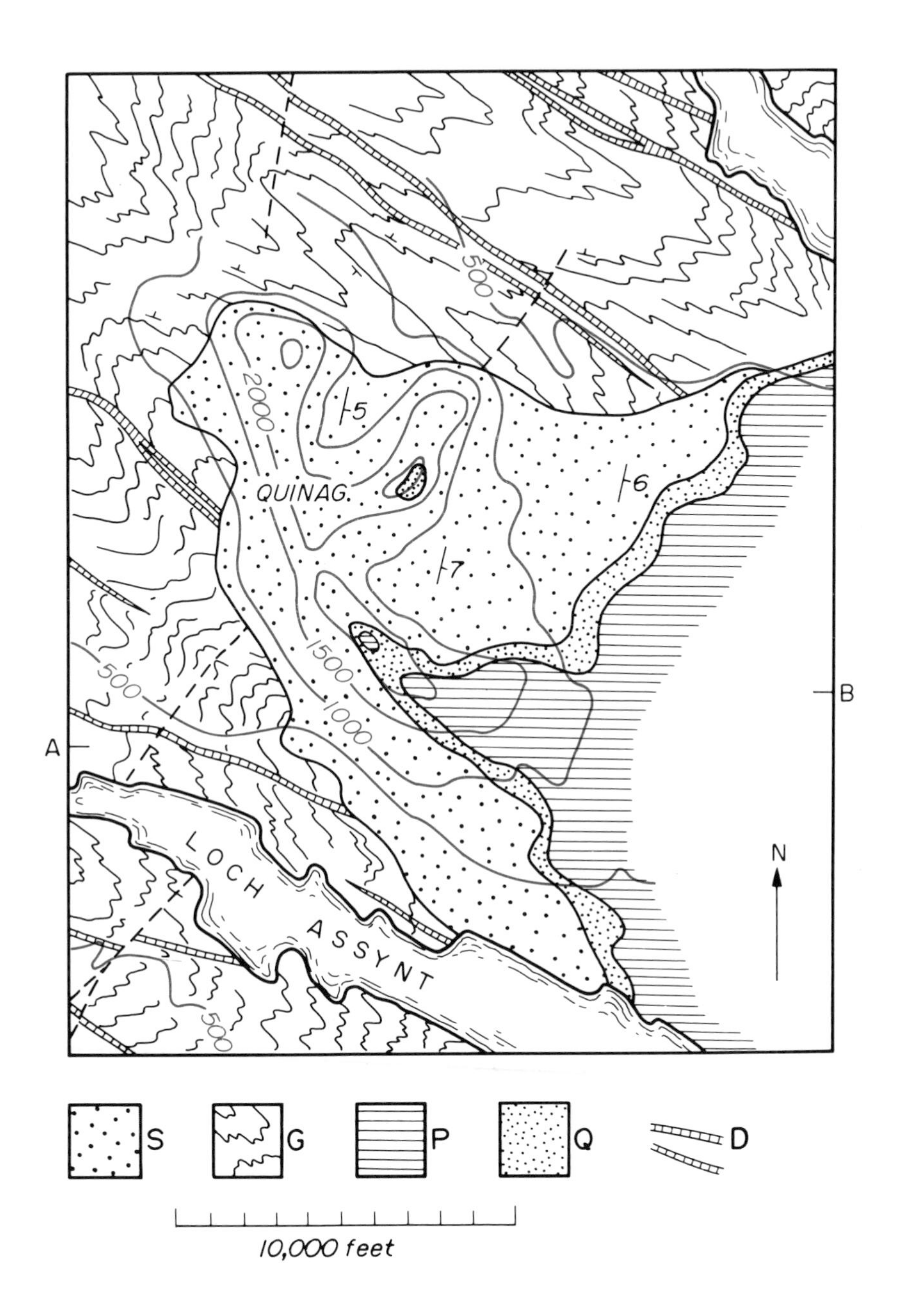

500
2000
5
QUINAG.
6
7
1500
1000
500
500
A
B
LOCH
ASSYNT
N
S
G
P
Q
D
10,000 feet

7

Folding with Cleavage

7.1 Foliations

In earlier chapters we have considered sedimentary rocks which possess a planar structure called bedding. The general name given to planar structures such as bedding is *foliation*. We say that bedding is a *primary foliation* because it is an original structure in the sediment and produced during sedimentation.

Other planar structures are produced in the rock at a later stage. During periods of earth movement, crustal rocks become subjected to stress (pressures which are not equal in all directions, but are greatest in one direction and least in a direction at right angles to the greatest pressure). Providing the rocks are ductile (are able to deform without fracturing) these stresses produce a permanent distortion or *strain* in the rocks.

The straining may be accompanied by a reorganization of the microscopic structure of the rock. Helped by a chemical alteration of the rocks, grains making up the rock may change shape and rotate so that their long dimensions turn away from the direction of greatest shortening in the rock. The new alignment of grains gives the rock a 'grain' or fabric. This structure is a type of *secondary foliation*.

The secondary foliation present in the finer-grained metamorphic rocks, such as slates, is called cleavage. *Slaty cleavage* usually gives a rock a well-developed ability to split into thin plates. Roofing slates, such as those of the Cambrian of North Wales, owe their fissility to the presence of slaty cleavage. Slaty cleavage is expressed on a microscopic scale by the alignment of grain shapes (Fig. 7.1A). From the shape of distorted (strained) fossils in cleaved rocks it can be demonstrated that slaty cleavage planes have formed perpendicular to the direction of greatest shortening. *Crenulation cleavage* is a foliation which is produced by a crinkling (small-scale folding, folds having a wavelength of about 1-10 mm) of pre-existing foliation. The crenulation cleavage planes are parallel to the axial planes of these microfolds (Fig. 7.2).

FIG. 7.1 **A**: Slaty cleavage. Electron microscope image (scale bar is 20 Mm long) of a Cambrian roofing slate to show the alignment of flaky minerals (mainly chlorite). This is typical of the microscopic structure of slaty cleavage. (Photograph: Prof. W. Davies.)

B: Slaty cleavage in the field. Bedding dips gently to the left; cleavage more steeply to the left of the photograph.

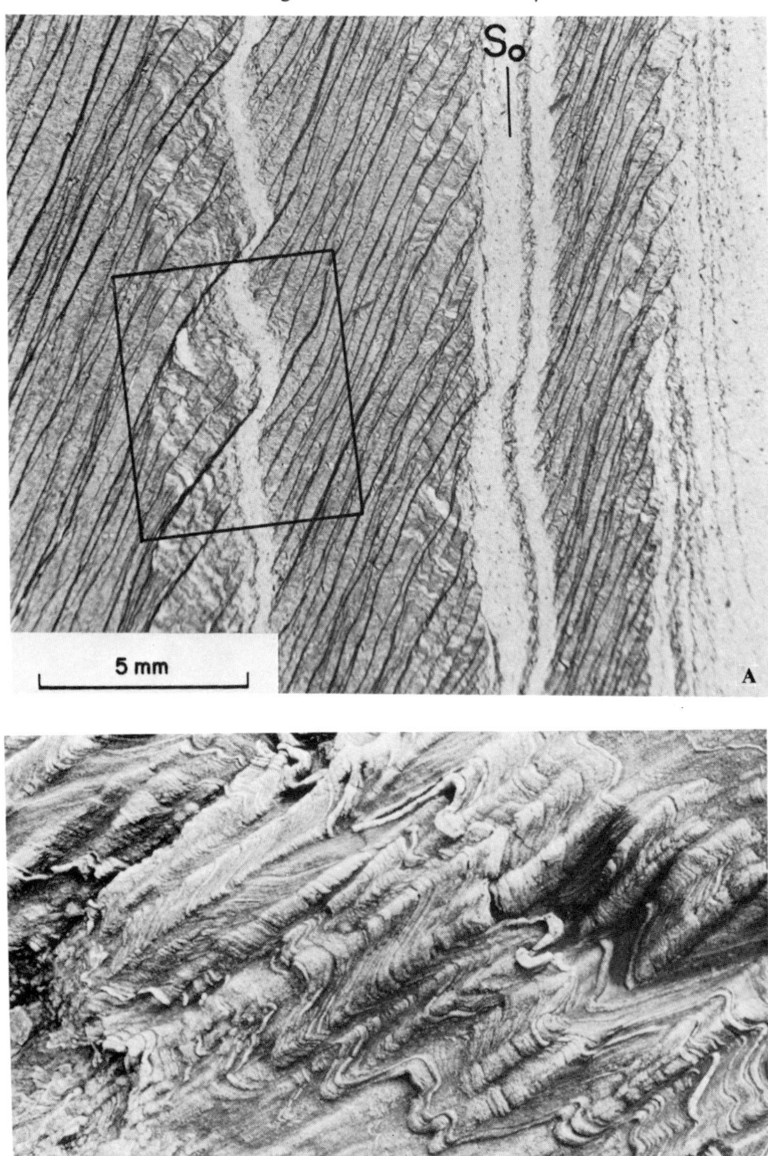

FIG. 7.2 Crenulation cleavage. **A**: The microscopic appearance of crenulation cleavage (set of planes sloping at 60° towards the left of the photograph). The crenulation cleavage is parallel to the axial planes of small folds which fold an earlier (older) foliation and bedding (appears vertical, labelled *So*), Tayvallich Slates, Oban, Scotland. (Photograph: Dr. G. Borradaile.) **B**: Crenulation cleavage (running diagonally across the photograph, top right to bottom left). Precambrian schists, Anglesey.

7.2 Axial Plane foliations

The strains needed to produce a cleavage in a rock are similar to those suffered by the rock's layers which have been forced to shorten and buckle to produce folds. Not surprisingly then, cleavage often occurs in rocks which are strongly folded. Cleavage and other foliations often have an attitude close to that of the axial plane (axial surface) of the fold (Figs 7.3, 7.4). The *axial plane cleavage* developed around a fold may not be perfectly parallel to the axial surface, but may vary in orientation across the fold (to define a *cleavage fan*) or from one bed to another (a feature called *cleavage refraction*, (Fig. 7.4B).

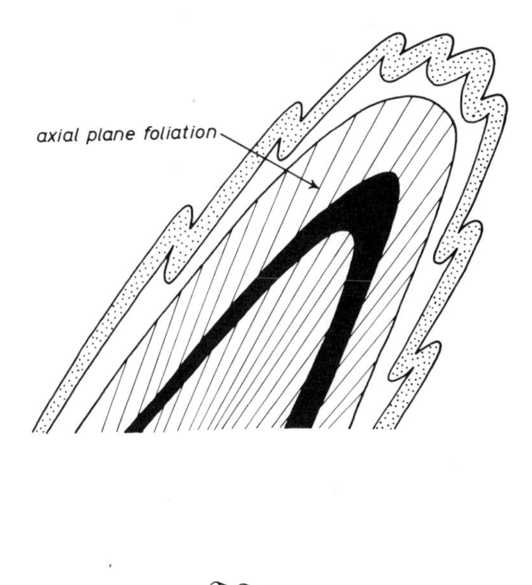

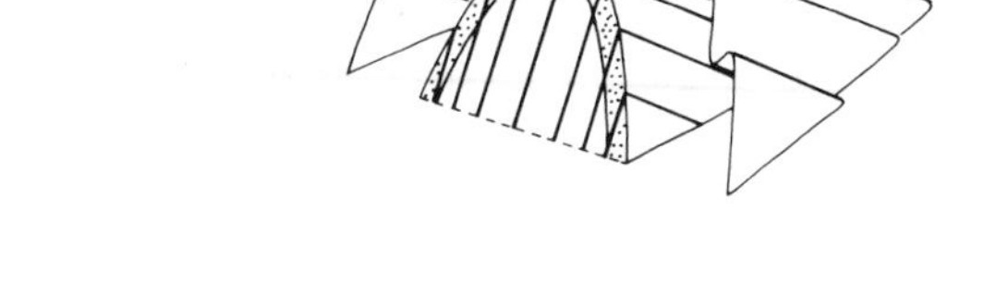

FIG. 7.3 The arrangement of small-scale folding and axial plane foliation (such as cleavage) in a larger-scale fold.

FIG. 7.4 Cleavage in relation to folds. The sense of obliquity between bedding and an axial plane cleavage at an outcrop can be used to interpret the position of the outcrop in relation to the larger-scale antiforms and synforms.

FIG. 7.4 **C**: Axial plane cleavage

7.3 The relationship of cleavage to bedding

An axial plane foliation in most instances has an attitude which is oblique to bedding (Figs 7.3, 7.4). An exception to this occurs in the hinge region of the fold where foliation and bedding are at right angles to each other. It can also be seen from Fig. 7.3 that the sense of obliquity between cleavage and bedding is different on each limb of the fold, i.e. these planes intersect one another in opposite senses. This fact is extremely important since it means that, by comparing cleavage and bedding attitudes at an outcrop, we are able to deduce our position with respect to large-scale folds in an area.

Worked example

Figure 7.5A shows a sketch of a rock outcrop where bedding and cleavage are visible. It is known that the bedding is folded in this area, and that the cleavage is axial planar to these folds. What can be deduced about these folds from the information at this outcrop?

We look first at the dip of the cleavage. This dips moderately to the left (west) and from this we should visualize folds occurring within the area with axial planes having the same dip. Some possible folds with westerly dipping axial planes are drawn in Fig. 7.5B. If we consider one such fold of anti-formal type, we find that the bedding/cleavage relationship on its east limb

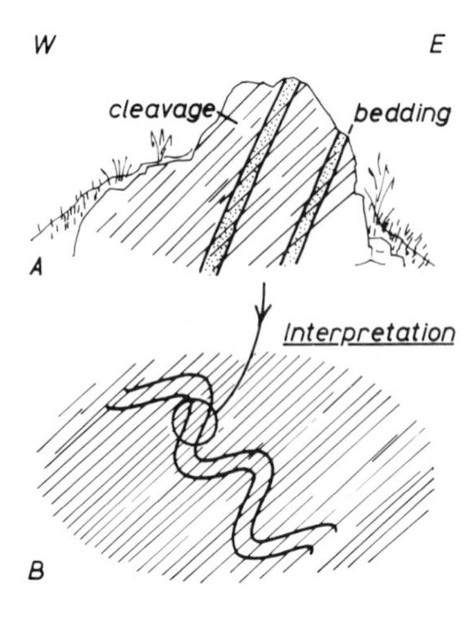

FIG. 7.5

matches that observed at the outcrop. We can conclude from this that the outcrop is situated on the east limb of an antiformal fold. In other words, we have deduced that the next fold hinge to the west of the outcrop is an antiform. We are, of course, not able to say how far to the west the antiform is.

This method of deciding where the next antiform is located is extremely useful in areas of tightly folded rocks, since in these areas the attitude of the limbs of the fold is not sufficiently different to allow each of the limbs to be distinguished on the basis of bedding dip alone.

With overturned folds, both limbs dip in roughly the same direction and on one of the limbs the beds are upside-down. Providing the folds are upward-facing (i.e. the antiforms are anticlines, the synforms are synclines), the overturned beds can be identified from the fact that on this limb the bedding will dip more steeply than the cleavage planes.

PROBLEM 7.1

A shows a vertical outcrop face at Ilfracombe, Devon, on which the traces of bedding and cleavage are visible. The cleavage, which exhibits refraction, is parallel to the hammer handle in that part of the outcrop (north is to the left of the photograph).

B shows a horizontal exposure at Tayvallich, Argyllshire, on which a normally graded sandstone bed crops out. The trace of cleavage is also visible and is parallel to the pen (north is to the left of the photograph).

Describe, with the aid of a sketch-map or cross-section, the possible larger-scale fold structures to which the structures at each of these outcrops belong.

PROBLEM 7.2

Interpret the structure from the bedding/cleavage data along the road section.

Describe the folds present. If these folds are upward-facing, locate the outcrops with inverted bedding.

PROBLEM 7.1

PROBLEM 7.2

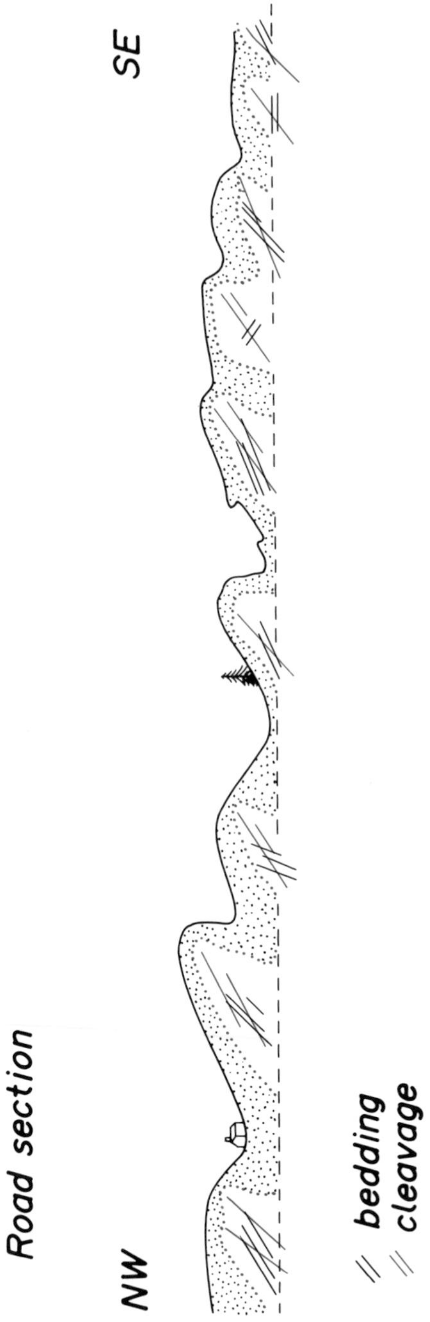

Road section

SE

NW

bedding

cleavage

Further Reading

(a) **Structural aspects of geological map interpretation**
J. L. Roberts *Introduction to Geological Maps and Structures*, 1982, Pergamon.
D. M. Ragan, *Structural Geology*, 1985. Wiley and Sons.

(b) **How geological maps are produced**
J. W. Barnes, *Basic Geological Mapping*, 1981. Open University Press.

(c) **Field techniques**
N. Fry, *The Field Description of Metamorphic Rocks*, 1984. Open University Press.
K. McClay, *The Mapping of Geological Structures,* 1987, Open University Press.

(d) **The geometry of structures and deformation of rocks**
J. G. Ramsay and M. Huber, *The Techniques of Modern Structural Geology*, 1987. Academic Press.
R. G. Park, *The Foundations of Structural Geology*, 1983. Blackie.

(e) **Foliations, fabrics and the study of structures under the microscope**
B. E. Hobbs, W. D. Means and P. F. Williams, *An Outline of Structural Geology*, 1976. Wiley and Sons.

Index